Chemistry of the Environment

E.N. Ramsden

B.Sc., Ph.D., D.Phil.

Formerly of Wolfreton School, Hull

Stanley Thornes (Publishers) Ltd.

First published 1996 by
Stanley Thornes (Publishers) Ltd
Ellenborough House
Wellington Street
CHELTENHAM
GL50 1YW

A catalogue record for this book is available from the British Library.

ISBN 0 7487 2400 1

99 00 / 10 9 8 7 6 5 4 3 2

The front cover shows a map of the distribution of ozone for the Southern Hemisphere in October 1993. The red areas show where the ozone concentration is high, and the blue area shows the 'ozone hole'. Normally the ozone concentration would be expected to be high over polar regions and relatively low over low latitudes (as indicated by the green colour). During the southern spring, as the map shows, around two-thirds of the total ozone is lost over Antarctica owing to chemical depletion attributable to CFCs. (Photograph reproduced by courtesy of NASA.)

Typeset by Tech-Set, Gateshead, Tyne & Wear.
Printed and bound in Great Britain by Redwood Books, Trowbridge Wiltshire.

CONTENTS

PREFACE

Chemistry of the Environment has been written to match the 1996 syllabuses for the following A-level modules:

University of Cambridge Local Examinations Syndicate:
    Chemistry 9535 Module 4824: Environmental Chemistry
    Chemistry 9254: Environmental Chemistry Option
University of Oxford Delegacy of Local Examinations Advanced Chemistry 9855
    Module 5 Option: Chemistry of the Environment
Oxford and Cambridge Schools Examination Board Chemistry Advanced Level 9684
    Unit C7 Environmental Chemistry

Before embarking on an optional topic such as *Chemistry of the Environment*, students will have completed the A-level Chemistry core modules, covering atomic structure, the chemical bond and a firm foundation of physical, inorganic and organic chemistry. Should they need to revise this core material, they can consult the references to *ALC* which are to my text, *A-level Chemistry,* Third edition (STP). They give the section of this text in which the relevant core material can be found. Students who are using a different A-level textbook need to consult the index of their book to find the corresponding material.

Acknowledgements

I offer my sincere thanks to Dr Rob Ritchie for his constructive comments on the first draft and for his kindness in completing his work so rapidly.

I am grateful to the following people and firms for supplying photographs:

British Nuclear Fuels plc: 8.16E
CEGB: 8.16F
David Hall/Panos Pictures: 3.10A
Environmental Picture Library: 5.1B (Stan Gamester); 5.3B (Air Images);
    8.8A (Steve Morgan); 8.9A (Charlotte Macpherson)
ICCE: 6.4A (Philip Steele); 6.11A (Sue Boulton)
ICI Chemicals and Polymers Ltd: 8.11A
NASA: 3.11A
Science Photo Library: 3.6B (Adam Hart-Davis); 4.16A, 8.7A (Martin Bond);
    5.4A (Shiela Terry); 6.11E (Simon Fraser)
Woodmansterne: 7.2B

I thank the publishing team who have contributed their enthusiasm and expertise to the production of this book and especially the Science Editor, John Hepburn, for his thoughtful work.

My family have given me their support and encouragement all through the writing of this book.

E.N. Ramsden
Oxford, 1996

1

THE ENVIRONMENT

Environmental science is traditionally divided into the study of the atmosphere, the hydrosphere, the geosphere and the biosphere.

1.1 THE ATMOSPHERE

The atmosphere is crucial in photosynthesis and respiration and in the water cycle.

The **atmosphere** is the thin layer of gases that covers the Earth's surface. It acts as a blanket to protect the Earth from receiving too much radiation from space and to stabilise the Earth's temperature. It provides carbon dioxide for photosynthesis, oxygen for respiration and nitrogen for incorporation into plant and animal proteins. The atmosphere plays its part in the water cycle, transporting water vapour from oceans to land.

1.1.1 THE EVOLUTION OF EARTH'S ATMOSPHERE

The atmosphere evolved from a mixture of volcanic gases.

The present-day atmosphere has evolved from a mixture of gases released from the interior of the Earth about 4.5 billion years ago. The mixture was probably similar to that released from volcanoes today: water vapour 64% by mass, carbon dioxide 24%, sulphur dioxide 10%, nitrogen 1.5%. This mixture has evolved into today's atmosphere which is mainly nitrogen and oxygen with only a trace of carbon dioxide.

As water vapour condensed, carbon dioxide dissolved in the oceans.

The Earth is far enough from the Sun to allow the primordial water vapour to condense to make oceans. Carbon dioxide dissolved in the oceans in large quantities and was eventually converted into limestone beds. Since Venus is closer to the Sun than the Earth is, oceans never formed there and carbon dioxide remains in the atmosphere.

Living things started to use it in photosynthesis which resulted in the production of oxygen.

When living things emerged on Earth, they altered the composition of the atmosphere. The first primitive living things evolved in the ocean about 3.5 billion years ago. The atmosphere was irradiated with ultraviolet (UV) light which provided the energy needed to bring about chemical reactions that resulted in the formation of substances such as amino acids and sugars. The first living things were aquatic. They derived energy by fermenting the organic matter that had been formed. Eventually they became able to carry out photosynthesis to produce organic matter (represented as CH_2O).

$$CO_2 + H_2O + h\nu \longrightarrow (CH_2O) + O_2$$

Organic
matter

1

The oxygen formed in photosynthesis would have poisoned primitive plants. It was not released into the air but was converted into iron oxides. Layers of iron oxides were laid down between 3 billion and 1.5 billion years ago. With no oxygen in the atmosphere, no ozone, O_3, was able to form. With no ozone layer in the atmosphere to absorb UV radiation from the Sun, the surface of the Earth was too hot to support life. However, bacteria were able to live in the ocean, protected from UV radiation by water. They eventually acquired enzymes that enabled them to use the oxygen in the atmosphere to oxidise organic matter in the ocean with the liberation of energy. The process evolved into respiration, the mechanism by which present-day organisms obtain energy.

Bacteria in the ocean evolved the ability to use oxygen in respiration.

Between 1800 and 800 million years ago, oxygen began to accumulate in the atmosphere. It enabled the ozone layer to form, shielding the Earth from excessive UV radiation and making it an environment in which living things could prosper (see §3.11). The Earth became a welcoming environment and land plants emerged 400 million years ago, followed by land animals 300 million years ago. Living things evolved from sea-dwellers into land-dwellers.

Oxygen accumulated in the atmosphere and the ozone layer formed, creating an environment in which living things could prosper on Earth.

Volcanic activity had released nitrogen since the beginning of Earth's existence. The nitrogen had built up since no living things had so far metabolised it. The composition of the atmosphere has remained the same for about 300 million years.

The atmosphere has had the same composition for 300 million years.

1.2 THE HYDROSPHERE

The **hydrosphere** contains the Earth's water. Over 97% of the Earth's water is in oceans. Water covers 70% of the Earth's surface. It occurs in all spheres of the environment: in the oceans as brine, as surface water in lakes and rivers, underground as groundwater, in the atmosphere as water vapour and in the polar ice caps as solid ice. Water is an essential part of all living things. Energy and matter are carried through the different spheres of the environment by water. Water dissolves out (leaches) soluble compounds from minerals and carries these solutes to the ocean or deposits them at some distance from the source. Water carries plant nutrients from soil into the bodies of plants through plant roots. Water absorbs solar energy as enthalpy of vaporisation when ocean water evaporates, and is carried inland and releases the energy when water vapour condenses.

Water is present in all spheres of the environment. It transfers matter and energy from one sphere to another.

1.3 THE GEOSPHERE

The **geosphere** is the part of the Earth on which human beings live and from which they extract food, fuel and minerals. The Earth consists of layers: a solid inner core, a molten outer core, a mantle and a crust. The outer mantle and the crust are called the **lithosphere**, which is 50–100 km in thickness. The lithosphere is directly involved with environmental processes. The thin outer layer, the **crust**, is 5–40 km thick and is the part which is accessible to human beings.

The geosphere is the part of the Earth on which human beings live.

Geology is the science of the geosphere. It is concerned mainly with the solid mineral portions of the Earth's crust. It also considers water, the atmosphere and climate, which have profound effects on the geosphere, and living things which exist on the geosphere. The most important part of the geosphere for life on Earth is **soil**, which has been formed by the weathering action of physical, geochemical and biological processes on rock. Soil is the medium on which plants grow and all land organisms depend on it for their existence. The productivity of soil is affected by environmental conditions and pollutants.

The crust is the thin outer layer of the Earth. The crust and the outer mantle make up the lithosphere. An important part of the geosphere is soil.

FIGURE 1.3A
The Layered Structure of
the Earth

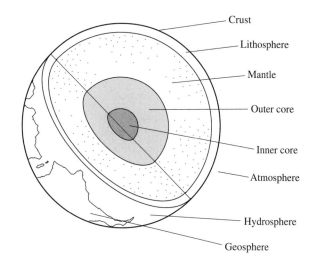

Crust

Lithosphere

Mantle

Outer core

Inner core

Atmosphere

Hydrosphere

Geosphere

1.4 THE BIOSPHERE

The biosphere is the total of all living things on Earth.

All living things on Earth make up the biosphere; they are the **biotic** part of the environment. Non-living parts of the environment are **abiotic**. The interactions between living organisms and the abiotic part of the environment are described by cycles that involve biological, chemical and geological processes. Examples are the carbon cycle [see § 2.8] and the water cycle [see § 4.1]. The study of the effects of chemicals in the environment on living things is **environmental biochemistry**.

1.5 ENVIRONMENTAL CHEMISTRY

Environmental chemistry is the study of the effects of chemicals on the environment.

Environmental chemistry is the study of the chemicals in water, soil and air: their sources, their reactions, their movement through the environment, their effects and the effects of technology on them. The study of environmental chemistry has grown rapidly since the beginning of the twentieth century. The effects of products and processes on the environment must now be considered when new techniques are under development. There are now many employment opportunities for environmental chemists. They contribute to industry by helping to develop processes which avoid pollution. Many times it is impossible to arrive at a simple answer to an environmental chemistry problem; many factors come into play, and the problems are not clear-cut. Sometimes long-term predictions of how environmental systems will behave must be made, and often compromise solutions have to be adopted.

There are many employment opportunities for environmental chemists ... ... who use a variety of analytical techniques.

One of the areas tackled by environmental chemists is the measurement of pollutants in the environment. **Analytical techniques** are therefore vital in environmental chemistry research. An air pollutant present at less that $1 \, \mu g \, m^{-3}$ of air may be significant. A water pollutant present at 1 ppm by mass may be a serious threat. Analytical methods used to study some environmental systems have a very low limit of detection.

In this book, chapters are devoted to the environmental chemistry of the atmosphere, the hydrosphere and the geosphere. Consideration of the effects of changes in the environment on the biosphere are woven into all these three spheres.

1.6 ENERGY

Two centuries ago, we relied on energy captured by photosynthesis: the energy of biomass as food to provide muscle power and as wood to provide heat. We have passed on to a dependency on fossil fuels – petroleum, natural gas and coal. The supplies of fossil fuels are limited, and their potential for pollution is high. The mining of coal and the extraction of petroleum disrupt the environment. The combustion of coal releases acidic sulphur dioxide into the atmosphere [§ 3.6]. All fossil fuels release carbon dioxide, a greenhouse gas [§ 2.9]. The move is towards alternative energy sources, especially renewable energy sources such as solar energy, and nuclear energy.

Our main source of energy is fossil fuels.

1.7 HUMAN IMPACT

The world population is increasing. Most people are aiming for a higher standard of living. The result is worldwide pollution on a massive scale – on land, in the air, in the hydrosphere. The three spheres are linked. Atmospheric pollutants, e.g. sulphur dioxide, fall to the earth in rain and pollute land and water. Discarded wastes which are dumped in landfill sites can leach into groundwater and become part of streams and rivers.

An increase in the use of fossil fuels had led to increased pollution.

1.8 POLLUTION

A **pollutant** is a substance which is present at a concentration greater than its natural concentration as a result of human activity and which has a detrimental effect on the environment. A **contaminant** is a substance which is present at a concentration greater than its natural concentration and is not classified as a pollutant unless it has a detrimental effect. Every pollutant originates from a **source**. It may act on a **receptor**, e.g. a human being. Eventually the pollutant may be deposited in a **sink** – a long-time repository – where it will remain for some time though probably not for ever.

The terms 'pollutant' and 'contaminant' are defined.

1.8.1 WATER POLLUTION

The quality of drinking water is a major factor in determining human welfare. Waterborne diseases in drinking water have killed millions of people. Industrialised nations have controlled waterborne diseases, but they are widespread in poorer countries where the populations are expanding and the resources needed to provide safe drinking water and to treat wastewater are not available. In industrialised nations the accidental discharge of toxic chemicals poses the greatest threat to the safety of water supplies.

Waterborne diseases and toxic waste threaten the safety of water supplies.

1.8.2 AIR POLLUTION

Inorganic air pollutants include several hundreds of millions of tonnes a year of oxides of carbon, sulphur and nitrogen. Carbon monoxide is toxic, carbon dioxide is a greenhouse gas, and oxides of sulphur and nitrogen contribute to acid rain. A **primary atmospheric pollutant** is a pollutant that is released directly into the atmosphere. A **secondary atmospheric pollutant** is one that is formed by reactions in the atmosphere. **Organic pollutants** may have a **direct effect**, e.g. cancer caused by exposure to chloroethene, and they may also contribute to the formation of **secondary pollutants**, e.g. photochemical smog. In the case of pollutant hydrocarbons in the atmosphere the formation of secondary pollutants is the important effect.

The terms 'primary pollutant' and 'secondary pollutant' are defined.

1.8.3 LAND POLLUTION

Disposal of wastes pollutes land.

Hazardous wastes may be discharged into the soil. A hazardous waste is a material that has been left where it may cause harm to human beings, animals and plants.

1.9 TECHNOLOGY

Human beings use materials and energy to make things. The methods that they use are called **technology**. We use technology to provide the food, housing and possessions that we need for our survival and for our comfort. Technology has a big influence on the environment. The challenge which environmental science faces is the cooperation of technology with considerations of the environment so that the two are mutually supportive and not in opposition. Technology can be applied for the protection of the environment, for example in improving manufacturing processes so as to minimise the production of hazardous waste products. Technology is being applied in the utilisation of renewable energy sources, e.g. solar power and wind power. In transport, technology increases the efficiency and safety of transporting people and goods.

We use technology to give us a good standard of life …

Some of the ways in which technology has altered the environment and caused pollution are:

… but technology can damage the environment …

- agricultural practices that enable intensive cultivation of land through the use of fertilisers, herbicides and insecticides

- processes for manufacturing industrial products that produce air pollutants, water pollutants and hazardous by-products

- energy production with environmental effects including disruption of soil by strip-mining and emission of air pollutants

- modern transport, with road construction, air pollution and increased demands for fossil fuels

Some of the ways in which technology can be applied to minimise environmental impact are:

… or, on the other hand, be used to protect the environment.

- computerised control of industrial processes to achieve optimum energy efficiency and minimum production of pollutants

- use of materials that minimise pollution problems, e.g. heat-resistant materials

- application of processes and materials that enable maximum recycling of materials and minimum waste production

- biotechnology, e.g. biological treatment of wastes

- use of the best catalysts

1. (*a*) How does Earth's atmosphere today differ from that of billions of years ago?

(*b*) How did living organisms alter the composition of the atmosphere?

2. Distinguish between the geosphere, the lithosphere and the Earth's crust.

3. Why has the science of environmental chemistry developed rapidly during the present century?

4. (*a*) Distinguish between a contaminant and a pollutant.

(*b*) What is meant by the source of a pollutant and its sink?

5. Mention some examples of how technology has been applied in an 'environmentally friendly' manner.

2

THE ATMOSPHERE

2.1 IMPORTANCE OF THE ATMOSPHERE

The atmosphere supports photosynthesis and respiration and provides nitrogen for fixation by plants and bacteria.

The mass of the atmosphere is 5.1×10^{15} tonnes. Of this total mass 99.99% is within 80 km of the surface of the Earth and 99% within 30 km . It forms a layer which accounts for 1% of the Earth's diameter. The atmosphere has many important functions.

1. It is a source of carbon dioxide for plant photosynthesis.

2. It is a source of oxygen for respiration.

3. It is a source of nitrogen for nitrogen-fixing plants and nitrogen-fixing bacteria.

4. As a part of the water cycle, the atmosphere transports water from oceans to land.

5. By absorbing most of the cosmic rays from outer space, the atmosphere protects organisms from their effects.

*The atmosphere transports water vapour . . .
. . . protects Earth from cosmic rays and from UV radiation . . .
. . . and stabilises Earth's temperature.*

6. It absorbs most of the electromagnetic radiation from the Sun, but transmits radiation in the visible, near-ultraviolet and near-infrared regions and radio waves, while filtering out ultraviolet radiation that would harm living organisms.

7. The Earth absorbs solar energy and re-emits it as infrared radiation. The atmosphere absorbs much of the infrared radiation, thus stabilising the Earth's temperature at a level which will support life.

2.2 WEATHER AND CLIMATE

Short-term changes in atmospheric conditions are described as 'weather' and long-term changes as 'climate'.

Variations in the state of the atmosphere over a short period of time are described as **weather**. The weather is defined in terms of temperature, clouds, winds, humidity, visibility (fog etc.), precipitation (snow, rain, etc.) and atmospheric pressure. Longer-term variations and trends in the factors that compose weather are described as **climate**.

Huge masses of air with different temperatures, pressures and moisture contents are separated by boundaries called **fronts**. The movement of masses of air that occurs horizontally is called **wind**. It takes place from regions of high pressure to regions of lower pressure. Vertically moving air is called an **air current**. As air moves upwards it expands, and this produces a cooling effect. The reverse happens when a cold mass of air at a high altitude sinks and contracts, becoming warmer [see Figure 2.2A].

Moving air masses constitute winds and air currents . . .

The movement of air masses transfers heat from ocean to land . A large fraction of solar energy falls on to the oceans, where it vaporises water. The movement of air masses carries water vapour from ocean to land. When water vapour later condenses from atmospheric air over the land heat is released, warming the land. The net result is a transfer of heat from ocean to land [see Figure 2.2A].

. . . and transport water vapour and heat from ocean to land.

FIGURE 2.2A
Movement of Air Masses

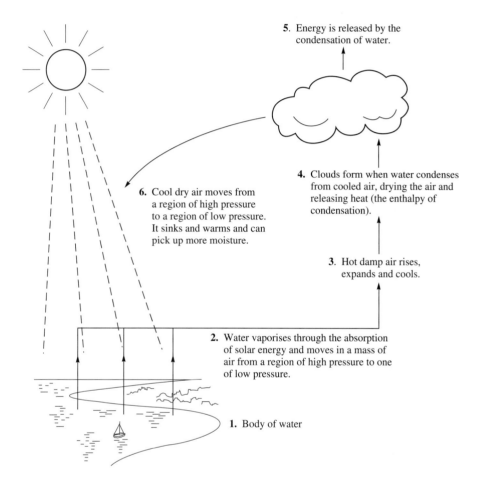

5. Energy is released by the condensation of water.

6. Cool dry air moves from a region of high pressure to a region of low pressure. It sinks and warms and can pick up more moisture.

4. Clouds form when water condenses from cooled air, drying the air and releasing heat (the enthalpy of condensation).

3. Hot damp air rises, expands and cools.

2. Water vaporises through the absorption of solar energy and moves in a mass of air from a region of high pressure to one of low pressure.

1. Body of water

The movement of air across Earth's surface is a crucial factor in the dispersal of air pollutants [see § 3.8].

2.3 STRATIFICATION

The temperature of the atmosphere varies with height and forms the basis for one classification of the atmosphere into layers as shown in Figure 2.3A.

The atmosphere can be divided on the basis of temperature into layers . . .

In the lowest layer of the atmosphere, the **troposphere**, the temperature decreases with altitude down to about $-56\,°C$. The composition of the troposphere is constant because there is constant mixing by circulating air masses. The water content is variable because of evaporation of water from oceans, cloud formation and precipitation. The low temperature of the **tropopause** – the top of the troposphere – is a barrier that makes water vapour condense to ice so that it does not reach altitudes at which it would receive high-energy UV radiation and undergo photolysis. If this happened the hydrogen produced would escape from the Earth's atmosphere.

. . . the troposphere, the lowest layer . . .

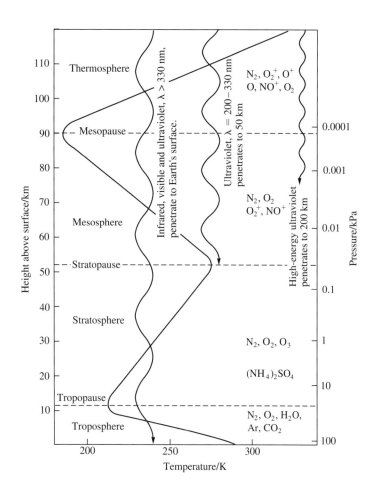

FIGURE 2.3A
Regions of the
Atmosphere

... the stratosphere ...

Directly above the troposphere is the **stratosphere**. The temperature rises with altitude in the stratosphere, reaching about −2 °C. The rise is caused by the absorption of UV light by ozone, which is present at about 10 ppm by volume in the middle of the stratosphere [see § 2.10].

... the mesosphere or chemosphere ...
... and the outermost thermosphere or ionosphere.

The **mesosphere** (sometimes called the **chemosphere**) has very low concentrations of species that can absorb radiation. The temperature falls to about −92 °C at an altitude of about 85 km. Extending to the far outer reaches of the atmosphere is the **thermosphere** (also called the **ionosphere**), which reaches temperatures as high as 1200 °C by the absorption of very energetic radiation of wavelength < 200 nm by atoms, radicals and molecules. Ions such as O_2^+, O^+ and NO^+ in the ionosphere reflect radio waves because of the charge they carry. Human beings have penetrated the mesosphere and thermosphere in spacecraft.

CHECKPOINT 2.3

1. Why does atmospheric pressure decrease with altitude?

2. A weather balloon is only partially inflated before it is released on its ascent into the stratosphere. Why is this?

3. Why does the temperature
(a) decrease with height in the troposphere

(b) increase above the tropopause
(c) fall in the mesosphere
(d) rise in the thermosphere?

4. Explain how the movement of a mass of air can transfer heat from sea to land.

2.4 PHOTOCHEMICAL REACTIONS

In **thermal reactions**, molecules are activated by heat to give them the **energy of activation** – the energy which colliding molecules must possess before a collision results in reaction [see *ALC*, § 14.9]. In **photochemical reactions**, molecules are activated by the absorption of light. Photochemical reactions depend on the production of free atoms or radicals. Figure 2.4A shows how the vibration, the distance between bonded atoms, depends on the potential energy of a diatomic molecule.

FIGURE 2.4A

The Potential Energy of a Diatomic Molecule

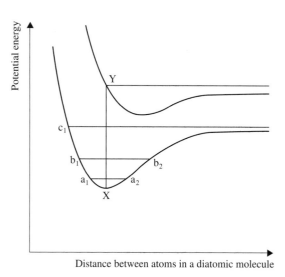

If a molecule possesses a small amount of vibrational energy it will vibrate so that the furthest distance between the atoms is $a_1 a_2$. With more energy it will vibrate between the limits $b_1 b_2$. If the vibrational energy is great enough, a vibration starting at c_1 will carry the atoms outside the range of the bonding forces and result in dissociation.

Suppose a molecule at point X in Figure 2.4A, with very little vibrational energy, absorbs a photon of light. The process is much more rapid than vibration: the interatomic distance does not change in the time it takes to absorb a photon. The molecule arrives at point Y. The excited state has its own potential energy curve. The energy of the excited state corresponds to a greater distance between the atoms than that of the unexcited state. By the time the molecule reaches the end of its vibration in the excited state, it will have fallen apart. The energy of a photon of light is given by

Photochemical reactions are important in atmospheric chemistry.

$$E = hv$$

where h = Planck's constant, v = frequency of light, $(= 1/\lambda$, λ = wavelength).

There is a frequency below which no reaction occurs because the energy of the photon must be great enough to cause dissociation.

Collision frequencies are low in the upper atmosphere, and the fraction of collisions that results in reaction must be high, as in free radical reactions.

Photochemical reactions play a big part in atmospheric chemistry. The absorption of light can bring about reactions which would not occur at the same temperature in the absence of light. The pressure is low in the upper atmosphere, the concentrations of reacting species are low, and collision frequencies are low. The fraction of collisions that results in reaction must be high if the reaction is to proceed. Free radical reactions are the only ones that can be effective in the upper atmosphere.

Ultraviolet radiation has a higher frequency than visible light and UV photons therefore have more energy and are more likely to break chemical bonds in the molecules that absorb them. For example, oxygen absorbs UV light of wavelength 135–76 nm and 240–60 nm in the stratosphere and molecules of O_2 dissociate to form two atoms of O. The atoms of oxygen formed each have an unpaired electron; they are free radicals. They rapidly react with O_2 molecules to form ozone, O_3.

UV light is more effective that visible light in forming free radicals.

(a) $O_2 + hv \, (UV) \longrightarrow 2O \cdot$

followed by exothermic reactions such as

(b) $O \cdot + O_3 \longrightarrow 2O_2; \quad \Delta H^{\ominus} = -390 \, \text{kJ mol}^{-1}$

(c) $O \cdot + O_2 + M \longrightarrow O_3 + M^*; \quad \Delta H^{\ominus} = -100 \, \text{kJ mol}^{-1}$

Reactions of free radicals often require the presence of a third body to absorb some of the energy released.

The reaction is very exothermic and a 'third body', M (possibly a molecule of nitrogen) must be present to absorb some of the energy released and prevent the dissociation of O_3. This limits the rate of reaction in the upper atmosphere where there is a low concentration of molecules that can act as third bodies. The energy acquired by M warms the stratosphere above the tropopause. The ozone that is formed absorbs UV radiation at wavelength 220–330 nm and causes the increase in temperature observed at the tropopause [see § 2.10 for the ozone layer]. Photochemical reactions such as those mentioned above do not occur at the surface of the Earth because the ozone which is formed in reaction (c) absorbs most of the UV photons having wavelengths < 340 nm.

The hydroxyl radical $\cdot$OH is the most important reactive intermediate in daytime photochemistry. Photolysis of oxygen gives excited oxygen atoms $O \cdot^*$. These react with water molecules to give hydroxyl radicals.

$$O \cdot^* + H_2O \longrightarrow 2 \cdot OH$$

The hydroxyl radicals react with oxygen atoms in a chain reaction.

Hydroxyl radicals are important intermediates.

$$\cdot OH + O \cdot \longrightarrow O_2 + H \cdot$$

$$H \cdot + O_3 \longrightarrow \cdot OH + O_2$$

These reactions play a part in removing ozone and maintaining the thickness of the ozone layer.

In night time photochemistry NO_2 radicals are the most important intermediates. The length of time that a chemical species is present in the atmosphere depends on the rate at which it enters the atmosphere from its **sources** and the rate at which it is removed to **sinks**. The source is the place where a substance originates. In the case of a pollutant the source is important because it is the logical place to eliminate pollution. A sink is a place where a species may be deposited and remain for a long time, though possibly not for ever. For example, a limestone wall may be a sink for atmospheric sulphuric acid. The lifetime of a chemical species is expressed as its **residence time**.

The length of time for which a species is present is expressed by its residence time.

$$\text{Residence time} = \frac{\text{Concentration of a given species in the atmosphere}}{\text{Rate of removal from the atmosphere}}$$

Nitrogen dioxide, NO_2, is one of the most photochemically active species found in a polluted atmosphere. A molecule of NO_2 absorbs light to form an electronically excited molecule.

$$NO_2 + hv \longrightarrow NO_2^*$$

NO_2 is an important intermediate . . . . . . with an unpaired electron which makes it chemically reactive.

A molecule may possess several excited states. The excited states have more energy than the ground state and are more chemically reactive. [For NO_2 in smog formation, see § 3.8.] The molecule NO_2 possesses an unpaired electron. The structure can be represented by combination of valence bond structures, such as (a) or by a delocalised π-bond structure as in (b).

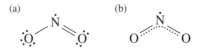

The unpaired electron explains why the reaction of a free radical with NO_2 is a chain-terminating reaction.

CHECKPOINT 2.4

1. What is meant by (*a*) a source,

(*b*) a sink,

(*c*) the residence time of a species in the atmosphere?

2. Why is the concentration of $\cdot OH$ radicals high in day time?

3. Why is nitrogen dioxide a reactive substance?

2.5 THE COMPOSITION OF THE TROPOSPHERE

The composition of the troposphere

The troposphere consists of nitrogen, oxygen, noble gases, carbon dioxide and traces of other gases. The presence of other gases depends on whether the air is sampled over an industrial city or a desert. The proportion of water vapour varies between 0% and 4%. The composition of unpolluted dry air is shown in Figure 2.5A.

FIGURE 2.5A
Composition of
Unpolluted Dry Air at Sea
Level

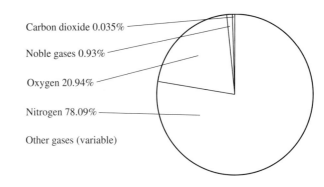

Carbon dioxide 0.035%

Noble gases 0.93%

Oxygen 20.94%

Nitrogen 78.09%

Other gases (variable)

2.6 THE NITROGEN CYCLE

Atmospheric nitrogen is fixed by some bacteria – converted into compounds which are used by plants in the synthesis of proteins. Animals obtain proteins from plants.

The atmosphere is an inexhaustible reservoir of nitrogen. Nitrogen is a very unreactive gas owing to the strength of the triple bond $N\equiv N$ (bond dissociation energy $945\,kJ\,mol^{-1}$). Neverthless nitrogen can be **fixed** – converted into compounds – by a number of methods. Bacteria in the root nodules of leguminous plants, e.g. peas and beans, contain the enzyme nitrogenase which catalyses fission of the $N\equiv N$ bond and enables the bacteria to convert nitrogen into soluble salts, which the plants can take in through their roots and use to synthesise plant proteins. Animals cannot synthesise proteins; they depend on eating plant proteins and the flesh of other animals.

Industrially, the fixation of nitrogen is achieved by the Haber Process [*ALC*, § 22.4.1], in which nitrogen and hydrogen combine at 400–600 °C and 100–150 atm to form ammonia.

$$N_2(g) + 3H_2(g) \rightleftharpoons 2NH_3(g)$$

The ammonia produced in the Haber Process is used in fertilisers as ammonia and as ammonium salts. It is also oxidised to nitric acid which has many uses, including the manufacture of nitrate fertilisers.

Some fixation of atmospheric nitrogen takes place during lightning storms with the formation of nitrogen monoxide, NO. This is slowly oxidised to nitrogen dioxide, NO_2, which dissolves in water in the atmosphere to form nitric and nitrous acids. Some of the nitrous acid is oxidised by oxygen in the air.

Other processes which fix nitrogen are the Haber Process, the formation of NO in vehicle engines and lightning storms.

$$N_2(g) + O_2(g) \longrightarrow 2NO(g)$$

$$2NO(g) + O_2(g) \longrightarrow 2NO_2(g)$$

$$2NO_2(g) + H_2O(l) \longrightarrow HNO_3(aq) + HNO_2(aq)$$

$$2HNO_2(aq) + O_2(g) \longrightarrow 2HNO_3(aq)$$

Rain carries nitric acid and nitrous acid to the soil. The same fixation of nitrogen takes place in the internal combustion engine, where the temperature is high enough for nitrogen and oxygen to combine to form nitrogen monoxide. This is present in the exhaust gases [see § 3.9].

2.6.1 NITRIFICATION

Ammonium compounds in the soil must be converted into nitrates before plants can use them. **Nitrification** is the oxidation of nitrogen(III) to nitrogen(V). It is important in soil and also in water.

$$2O_2(aq) + NH_4^+(aq) \longrightarrow NO_3^-(aq) + 2H^+(aq) + H_2O$$

Nitrification is the oxidation of ammonium compounds to nitrates ... ... catalysed by aerobic bacteria.

In Nature, nitrification is catalysed by bacteria. *Nitrosomonas* catalyses the oxidation of ammonium ion to nitrite ion.

$$2NH_4^+(aq) + 3O_2(aq) \longrightarrow 4H^+(aq) + 2NO_2^-(aq) + 2H_2O(l)$$

Nitrobacter catalyses the oxidation of nitrite ion to nitrate ion.

$$2NO_2^-(aq) + O_2(aq) \longrightarrow 2NO_3^-(aq)$$

Both bacteria function only in the presence of molecular oxygen. In waterlogged soils, the oxygen content may become too low and nitrification may cease [see § 7.11].

2.6.2 DENITRIFICATION

Nitrogen is lost from the soil in the process of **denitrification**. It occurs when the oxygen content of soil is low and anaerobic bacteria reduce nitrate ions. It is the natural mechanism by which fixed nitrogen is returned to the atmosphere. The reduction product of denitrification is a nitrogen-containing gas, usually nitrogen, but nitrogen oxides may also be produced.

Denitrification is the reduction of nitrates, catalysed by anaerobic bacteria.

$$4NO_3^-(aq) + 5(CH_2O) + 4H^+(aq) \longrightarrow 2N_2(g) + 5CO_2(g) + 7H_2O(l)$$

$(CH_2O) = $ Organic matter

The natural processes which remove nitrogen from the atmosphere and those which return nitrogen to the atmosphere are in equilibrium:

Rate of removal of nitrogen = Rate of return of nitrogen

The balance between processes which remove nitrogen from the atmosphere and those that return it is called the nitrogen cycle.

So well do the various reactions respond to the concentration of atmospheric nitrogen that the composition of the atmosphere has remained the same for 300 million years. Human activities interfere with the natural cycle. Crops are harvested instead of being left to die and decay, returning nitrogen compounds to the soil. Excreta are not returned to the soil. To make up for this, synthetic fertilisers are added to replenish the nitrogen content of the soil. The movement of nitrogen through the atmosphere and the lithosphere is called the **nitrogen cycle** [see Figure 2.6A].

FIGURE 2.6A
The Nitrogen Cycle

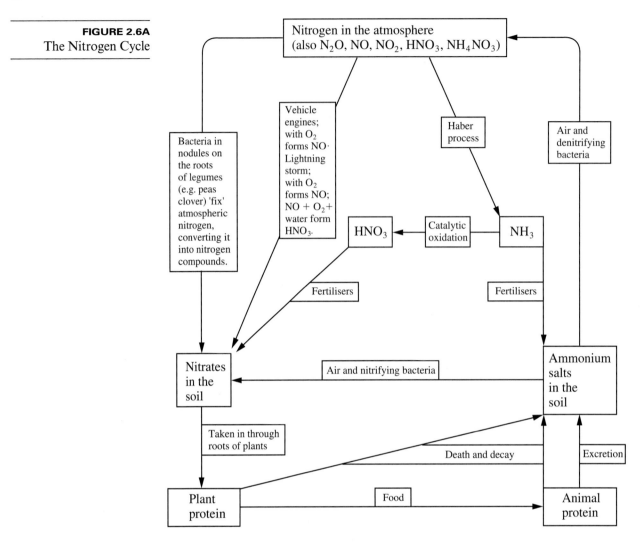

2.7 THE OXYGEN CYCLE

Oxygen is essential for the aerobic respiration of plants and animals.

$$C_6H_{12}O_6(aq) + 6O_2(g) \longrightarrow 6CO_2(g) + 6H_2O(l); \quad \Delta H^{\ominus} = -2820\,\text{kJ mol}^{-1}$$
Glucose

The energy released enables living processes to take place in plants and animals. Carbon dioxide and water are produced. In anaerobic respiration, in the absence of air, the process is:

$$C_6H_{12}O_6(aq) \longrightarrow 3CH_4(g) + 3CO_2(g)$$

and less energy is released.

During the photosynthesis of plants oxygen is returned to the atmosphere.

$$6CO_2(g) + 6H_2O(l) \longrightarrow C_6H_{12}O_6(aq) + 6O_2(g); \quad \Delta H^{\ominus} = +2820\,\text{kJ mol}^{-1}$$
Glucose

The balance between processes which remove oxygen, e.g. respiration, combustion and formation of ozone ...

The equilibria that exist between the processes which remove oxygen from the atmosphere and those that return it ensure that:

Rate of removal of oxygen = Rate of return of oxygen

The percentage of oxygen in the atmosphere remains constant as a result of these equilibria. The way in which oxygen passes from the atmosphere into compounds and back again is called the **oxygen cycle** [see Figure 2.7A].

FIGURE 2.7A

The Oxygen Cycle

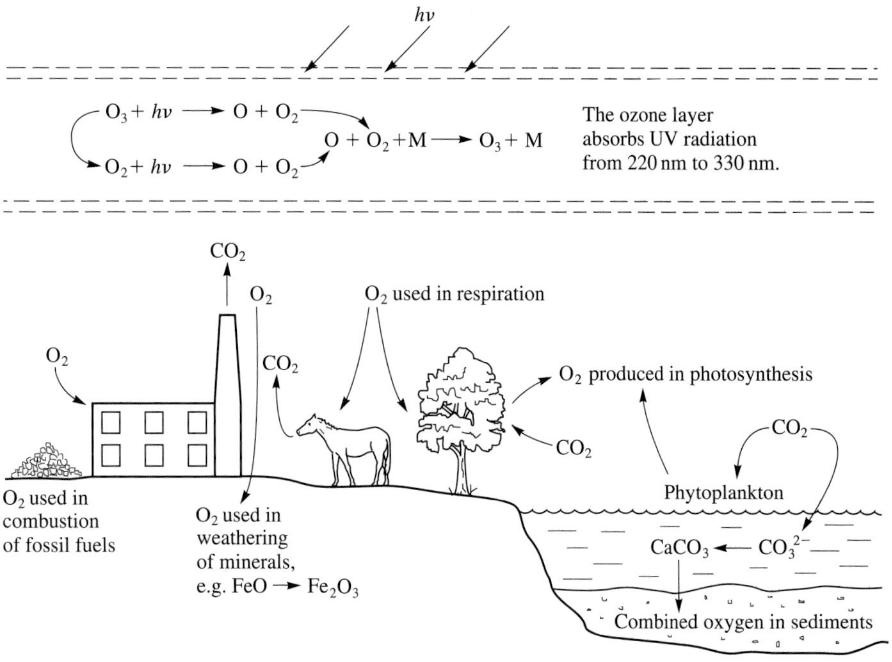

... and those that return oxygen, e.g. photosynthesis and decomposition of ozone ...
... is called the oxygen cycle.

In the upper atmosphere, ionising radiation converts oxygen into forms other than O_2. The stratosphere contains oxygen atoms, $O\cdot$, excited oxygen molecules, O_2^*, and ozone, O_3.

In the thermosphere (or ionosphere), atomic oxygen is produced by the photochemical reaction:

$$O_2 + h\nu \longrightarrow 2O\cdot$$

which can be followed by the formation of O^+:

$$O\cdot + h\nu \longrightarrow O^+ + e^-$$

and O^+ can react with molecular oxygen and nitrogen to form other ions:

$$O^+ + O_2 \longrightarrow O_2^+ + O$$

$$O^+ + N_2 \longrightarrow NO^+ + N$$

2.7.1 ARE WE GOING TO RUN OUT OF OXYGEN?

In the combustion of fossil fuels, for every molecule of CO_2 formed, a molecule of O_2 is consumed. In total a mass of 18 billion tonnes (18×10^9 tonnes) of atmospheric oxygen is consumed per year. Fortunately this is trivial compared with the mass that surrounds the Earth: 1×10^{15} tonnes. It would take 2000 years at this rate for the level of oxygen in the atmosphere to fall from 21% to 20%.

The atmosphere holds a vast supply of oxygen.

2.8 THE CARBON CYCLE

The carbon dioxide content of air is only 0.035% by volume. It is maintained at this level by a balance between processes which remove carbon dioxide from the atmosphere and those that return it –called the **carbon cycle** [see Figure 2. 8A].

FIGURE 2.8A
The Carbon Cycle

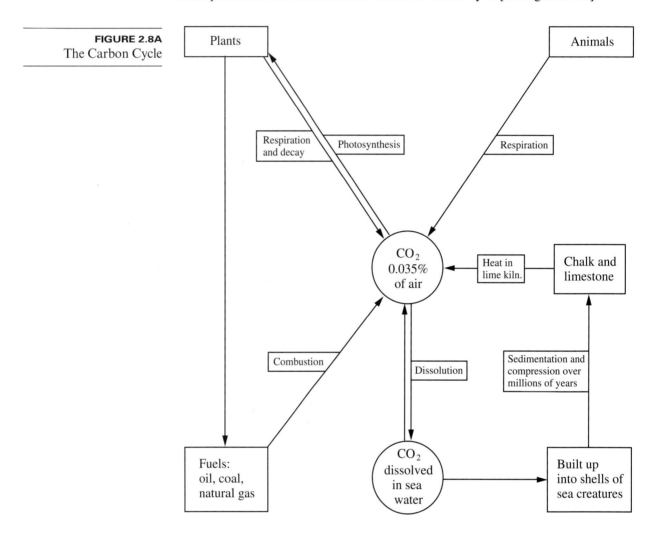

Carbon dioxide is essential for photosynthesis in plants.

Carbon dioxide is used in photosynthesis ...

sunlight, chlorophyll

$$6CO_2(g) + 6H_2O(l) \xrightarrow[\text{chlorophyll}]{\text{sunlight}} C_6H_{12}O_6(aq) + 6O_2(g)$$
$$\text{Glucose}$$

... and produced in respiration.

In respiration of plants and animals, carbon dioxide is produced.

$$C_6H_{12}O_6(aq) + 6O_2(g) \longrightarrow 6CO_2(g) + 6H_2O(l)$$
$$\text{Glucose}$$

The combustion of fossil fuels sends carbon dioxide into the atmosphere.

$$CH_4(g) + 2O_2(g) \longrightarrow CO_2(g) + 2H_2O(l)$$

The biodegradation of dead plant and animal material releases carbon dioxide into the atmosphere.

Much carbon dioxide dissolves in the oceans.

Carbon dioxide dissolves in water and equilibrium exists between gaseous carbon dioxide and aqueous carbon dioxide.

$$CO_2(g) \rightleftharpoons CO_2(aq)$$

When the concentration of carbon dioxide in the atmosphere increases, the position of equilibrium moves from left to right, and more carbon dioxide dissolves. If the level of atmospheric carbon dioxide falls, carbon dioxide can be released from the oceans to restore the level. The oceans play a major role in maintaining the level of carbon dioxide. Human activites, such as the combustion of fossil fuels, release an enormous volume of carbon dioxide into the atmosphere. The oceans have so far managed to dissolve it with only a small increase in atmospheric carbon dioxide. The implications of the small but steady increase in atmospheric carbon dioxide are discussed in §2.9.

The balance between processes which remove carbon dioxode from the atmosphere, e.g. photosynthesis and dissolution ...

Carbon dioxide reacts with water to form carbonic acid. This is a weak acid, and the equilibria set up are:

$$CO_2(aq) + 2H_2O(l) \rightleftharpoons H_3O^+(aq) + HCO_3^-(aq)$$

$$HCO_3^-(aq) + H_2O(l) \rightleftharpoons H_3O^+(aq) + CO_3^{2-}(aq)$$

... and those which send carbon dioxide into the atmosphere, e.g. respiration, combustion and release from oceans ...

The hydrogencarbonate ion is dissociated to a very small extent and the concentration of carbonate ion is very low [see §4.15.5].

... is called the carbon cycle.

The carbon dioxide content of the atmosphere is higher at night than by day when plants are photosynthesising. It is also affected by the seasonal variations that occur in the uptake of carbon dioxide by plants [see Figure 2.7A]. In the northern hemisphere maximum levels of carbon dioxide occur in April and minimum values in October. During the summer months forests carry out enough photosynthesis to reduce the atmospheric carbon dioxide content. During the winter, processes such as the bacterial decay of humus release carbon dioxide.

Carbon dioxide levels vary with the time of day and with the season.

There is a worldwide trend for the destruction of forests. Many countries in South America and Africa are felling their tropical rainforests to provide more land to graze cattle and grow crops. Tropical rainforests are warm and humid, conditions which are ideal for photosynthesis. Their destruction is one of the factors that is causing an increase in atmospheric carbon dioxide levels [see §2.9].

FIGURE 2.8B
A Tropical Rainforest

================================ CHECKPOINT 2.8 ================================

1. How do humans interfere with the natural nitrogen cycle with respect to:

(*a*) converting nitrogen into nitrogen compounds

(*b*) removing nitrogen compounds from the cycle

(*c*) supplying nitrogen compounds to the cycle?

2. What is the difference between aerobic respiration and anaerobic respiration?

3. Give the formulae of three oxygen species present in the stratosphere.

4. How could oxygen combined in a sediment of calcium carbonate [Figure 2.5A] become part of atmospheric oxygen?

5. How do bacteria contribute to the nitrogen cycle?

6. (*a*) How does the carbon cycle tie in with the oxygen cycle?

(*b*) Why does the level of atmospheric carbon dioxide vary from morning to night and from spring to autumn?

2.9 THE GREENHOUSE EFFECT

Earth's temperature is maintained at an average of 15 °C. One mechanism for doing this is the greenhouse effect.

The temperature of the Earth is fixed by a steady-state balance between the energy received from the Sun and an equal quantity of energy radiated back into space by the Earth. If anything were to upset this balance, the average temperature of the Earth would drift to a different steady-state value. The resulting changes in the Earth's climate could create deserts, melt the ice caps and raise the level of the oceans or start a new ice age. A number of factors maintain the Earth's temperature within very narrow limits at an average of 15 °C and maintain a climate that will support life. One mechanism for regulating the Earth's temperature is the **greenhouse effect**.

Energy is received by the Earth from the Sun and is radiated back into space.

The solar energy that reaches the troposphere is largely in the visible region of the spectrum [see Figure 2.9A]. It amounts to 1.34×10^3 watts per square metre of the Earth's surface ($80.6 \, kJ \, m^{-2}$), and if all this energy reached the Earth's surface, the Earth would be too hot to support life. In fact, about half of the solar radiation which enters the atmosphere reaches the Earth's surface. The other half is either reflected directly back into space or is absorbed in the atmosphere and its energy radiated back into space as infrared radiation. Most of the solar energy reaching the Earth's surface is absorbed; it must be returned into space to maintain the Earth's temperature.

Energy is lost from the Earth by conduction, convection and radiation. Infrared radiation returns energy into space.

The loss of energy from the Earth is achieved by means of conduction, convection and radiation. A fraction of the Earth's heat is transported to clouds by conduction and convection before being lost by radiation. Convection carries heat in the form of the enthalpy of vaporisation of water vapour which releases heat as it condenses. Radiation can transmit energy through a vacuum, and it is by means of radiation that all the energy lost by the planet is returned into space. The radiation that carries energy away from the Earth is of longer wavelength, in the infrared (IR) region, than the sunlight that brings energy to the Earth.

Water vapour, carbon dioxide and other gases absorb much of the IR radiation …

If all the outgoing radiation were able to escape the surface of the Earth would be at $-18 \, °C$, the same temperature as that of the Moon. However, water vapour, carbon dioxide and other gases absorb much of the outgoing radiation and re-radiate about half of it back to the Earth's surface. Most of the absorption is done by water molecules in the atmosphere. Absorption is weak at 7–8.5 μm and 11–14 μm and non-existent at 8.5–11 μm, leaving a 'hole' in the IR absorption spectrum through which radiation may escape. Carbon dioxide is present at lower concentrations than water vapour but absorbs strongly at 12–16.3 μm and therefore plays a key role in the greenhouse effect. An increase in the carbon dioxide level of the atmosphere could prevent sufficient energy loss and cause a damaging increase in the Earth's temperature. A decrease in the atmospheric level of carbon dioxide would cause a fall in the Earth's temperature. Other greenhouse gases are methane, dinitrogen oxide, carbon monoxide, ozone and CFCs.

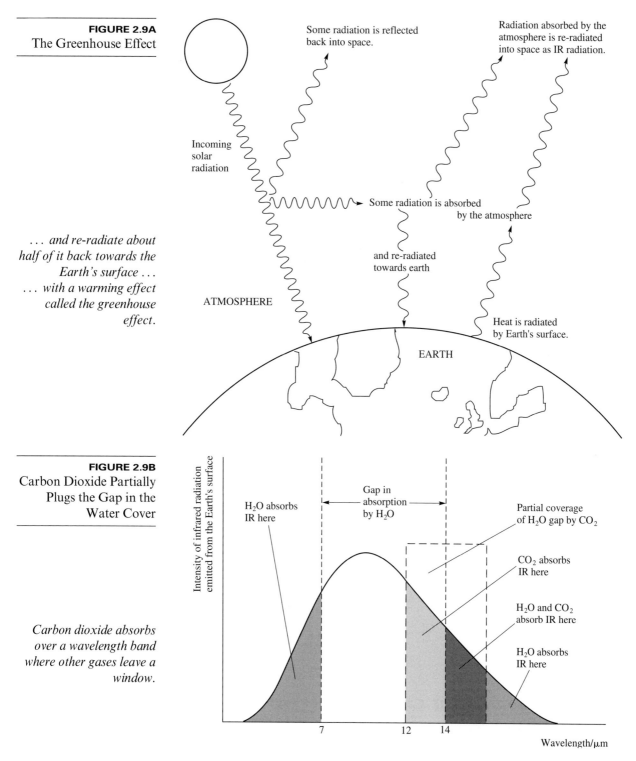

FIGURE 2.9A
The Greenhouse Effect

Some radiation is reflected back into space.

Radiation absorbed by the atmosphere is re-radiated into space as IR radiation.

Incoming solar radiation

... and re-radiate about half of it back towards the Earth's surface ...
... with a warming effect called the greenhouse effect.

Some radiation is absorbed by the atmosphere

and re-radiated towards earth

ATMOSPHERE

Heat is radiated by Earth's surface.

EARTH

FIGURE 2.9B
Carbon Dioxide Partially Plugs the Gap in the Water Cover

Carbon dioxide absorbs over a wavelength band where other gases leave a window.

Intensity of infrared radiation emitted from the Earth's surface

H_2O absorbs IR here

Gap in absorption by H_2O

Partial coverage of H_2O gap by CO_2

CO_2 absorbs IR here

H_2O and CO_2 absorb IR here

H_2O absorbs IR here

7   12   14

Wavelength/μm

2.9.1 IS THE CONCENTRATION OF ATMOSPHERIC CARBON DIOXIDE INCREASING?

Human activities generate carbon dioxide in six major ways. Three methods which return carbon dioxide to the air are:

- breathing   1 billion tonnes per year
- rearing farm animals   1 billion tonnes per year
- burning wood   7 billion tonnes per year

Human activities generate carbon dioxide: about 32 billion tonnes a year.

Three methods which add carbon dioxide to the atmosphere from sources that were underground are:

- making cement and lime from limestone   0.5 billion tonnes per year
- burning fossil fuels   22 billlion tonnes per year
- draining land for agricultural use   difficult to estimate

Most of this is absorbed by plant photosynthesis and by dissolution in oceans.

The total is about 32 billion tonnes a year, 6 tonnes for every person on the planet. The carbon dioxide in the atmosphere rises by only a fraction of this huge emission. The major part dissolves in the oceans and in rainwater and is photosynthesised by land plants and by phytoplankton in the upper layers of the oceans [see the carbon cycle, Figure 2.8A]. The atmospheric concentration of carbon dioxide can be measured spectrophotometrically. Meaningful measurements of carbon dioxide concentration can be taken only in regions remote from industrial activity. Elsewhere the combustion of fossil fuels raises the local level of carbon dioxide. Measurements made in Antarctica and the top of Mauna Loa Mountain in Hawaii over the past 30 years suggest that an increase of 1 ppm per year in the level of carbon dioxide has occurred [see Figure 2.9C]. The increase in carbon dioxide could come from the increased combustion of fossil fuels. Another factor is the destruction of vast areas of forest for agricultural purposes, which means that there are fewer trees to take in carbon dioxide in photosynthesis [see the carbon cycle, §2.8].

Measurements show that the concentration of carbon dioxide in the atmosphere is increasing.

FIGURE 2.9C
Atmospheric Carbon
Dioxide Recorded at
Mauna Loa, Hawaii

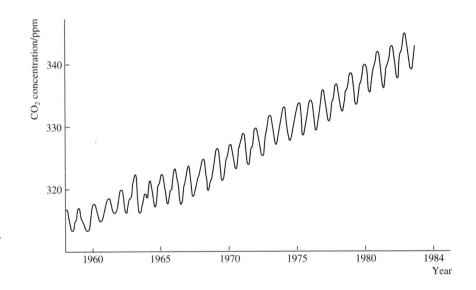

One factor is the increased combustion of fossil fuels. Another factor is the destruction of vast areas of forests.

2.9.2   IS THE EARTH'S TEMPERATURE RISING?

The Earth's temperature has risen over the past 140 years. This is a small rise compared with past fluctuations in Earth's temperature.

There is an urgent need to find out whether the Earth's temperature is rising, that is, whether **global warming** is taking place. Temperatures are recorded all over the Earth, but it is not easy to obtain an average value because few readings are taken at sea. In 1993 the National Aeronautics and Space Administration, NASA, reported the results of a 14-year study. Using a satellite to record average temperatures over the whole planet, they found no global warming between 1979 and 1993. At present it is very difficult to decide whether serious global warming is taking place. There are always small variations in temperature about a mean value. It is necessary to follow the temperature for a long time in order to establish a trend. Figure 2.9D shows an upward trend over the past 140 years. This is a small rise compared with fluctuations in temperature that have occurred in the past.

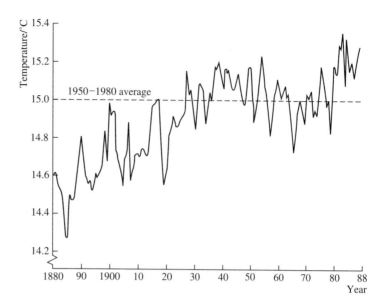

FIGURE 2.9D
Global Temperature over 140 Years

Estimates of Earth's temperature in the past are made by sampling Greenland ice at different depths.

To find out what the Earth's temperature was in the past, the Greenland Ice Core Project sampled snow that had fallen over the past 260 000 years. A research group drilled into the polar ice cap to analyse ice samples from different depths. The ratio of $^{18}O/^{16}O$ gives a measure of the temperature at which the ice formed: the lower the ratio of $^{18}O/^{16}O$, the colder the climate. (More energy is needed to vaporise $H_2{}^{18}O$ than $H_2{}^{16}O$). They found that there have been two Ice Ages of 90 000 and 100 000 years and three warm periods, including the present one, which has lasted for 11 000 years so far. During the previous warm period, the temperature of the Earth fluctuated. In the warmest years the temperature was 4 °C higher than at present; in the coldest years the temperature fell by 10 °C. The Earth remained cold for decades or centuries before warming again. The temperature increase shown in Figure 2.9D is small compared with these past variations. However, if we wait for a century before deciding whether it is a serious threat, we may have waited too long to avoid serious consequences. The urgent need is for more accurate temperature measurements to establish whether the trend is long-term.

If the upward trend continues for long, serious global warming will result.

2.9.3 IS THE RISE IN EARTH'S TEMPERATURE RELATED TO CARBON DIOXIDE?

There seems to be no doubt that the level of carbon dioxide in the atmosphere is rising. Whether it is the cause of the increase in temperature is more open to question. Scientists have investigated whether high levels of atmospheric carbon dioxide have been associated with a rise in global temperature in the past by analysing polar ice as described in §2.9.2. The first group to report their results caused a great deal of concern by reporting that the concentration of carbon dioxide was lowest in the ice laid down during the coldest periods. They inferred that a raised level of carbon dioxide would result in global warming. However a second group of scientists claimed in 1992 that their improved methods of sampling and more accurate analysis contradicted the previous results. So far, the connection between carbon dioxide levels and global warming is not established, and more accurate measurements are needed.

The question of whether the rise in Earth's temperature is caused by a rise in the level of atmospheric carbon dioxide is the subject of research.

2.9.4 METHANE

Methane is an important greenhouse gas. It is produced in large quantities by living organisms. There is a common group of bacteria called **methanogenic bacteria** which respire anaerobically to convert carbohydrates into methane and other compounds. The reaction is not simple, but a generalised equation is:

$$2(CH_2O) \longrightarrow CH_4(g) + CO_2(g)$$
Organic
matter

Methane is a greenhouse gas. It is produced by the anaerobic decay of vegetation...

Methane is produced whenever material containing carbohydrate is left in anaerobic conditions. Some examples are:

● In compost heaps and waste tips, vegetation rots in the absence of air and methane is formed. In biogas digesters the methane is used as a fuel.

● In the alimentary canals of animals, partly digested food is acted on by bacteria to form methane. (One cow releases about 500 dm^3 of methane a day!)

● Natural gas, which is mainly methane, was formed geological ages ago by the anaerobic decay of vegetation.

... e.g. in compost heaps, in rice paddies and in the guts of animals.

● In rice paddies rotting vegetation is covered by water and mud and so decays anaerobically to form methane.

The level of methane in the atmosphere is rising. When water freezes, bubbles of air become trapped in the ice. When ice forms at the North and South Poles, bubbles of air become trapped in polar ice and remain there for centuries. By analysing bubbles of air in polar ice of different ages, a picture of the methane content of the air at different times in the past can be obtained [see Figure 2.9E]. The increase in methane concentration over the last century has been much greater than in the preceeding 3000 years.

FIGURE 2.9E
Methane Concentration
in Polar Ice

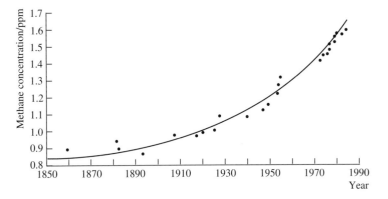

Methane and ozone are the most serious greenhouse gases, other than carbon dioxide. Their levels are rising at 1% and 0.5% a year respectively. Although there are natural routes for destroying methane in the atmosphere, these processes generate ozone at the same time as destroying methane. Methane in the atmosphere is attacked by hydroxyl radicals, HO·, to form methyl radicals, ·CH$_3$, which enter a chain of reactions that terminate in ozone. The oxidation of methane in the atmosphere thus results in the formation of ozone. The train of events is:

Methane contributes to the formation of ozone ...

$$CH_4 \xrightarrow{\cdot OH} CH_3\cdot \xrightarrow{O_2} CH_3OO\cdot \longrightarrow HOO\cdot \xrightarrow{NO} HO\cdot + NO_2 \longrightarrow NO\cdot + O\cdot \xrightarrow{O_2} O_3$$

| Methane | Methyl radical | Methylperoxy radical | Hydroperoxy radical | Hydroxyl radical | Nitrogen monoxide | Ozone |

. . . and to the formation of photochemical smog.

In addition to being a greenhouse gas, ozone is a pollutant in the troposphere and a contributor to photochemical smog [§ 3.8]. If governments decide that the evidence shows that global warming is occurring, the emissions of both carbon dioxide and methane must be reduced.

2.9.5 THE INTERGOVERNMENTAL PANEL ON CLIMATE CHANGE

The Intergovernmental Panel on Climate Change asked in 1990 for an immediate reduction in the emission of greenhouse gases.

The Intergovernmental Panel on Climate Change was set up in 1988. In 1990 a report from the majority of the 170 scientists in 25 countries – with the exception of some scientists who dissented – asked for an immediate reduction in the emission of greenhouse gases. The Panel predicted that if emissions continue at the present levels the temperature of the Earth will rise by 3 °C in the next century and the sea level will rise by 65 cm. The report admitted that there were uncertainties because the computer simulations on which it was based did not take into account clouds, oceans and polar ice sheets. The predictions led to a great deal of alarm about global warming, with forecasts of the sea rising by up to 9 metres. The predictions calculated that as the temperature rose more water would vaporise from the oceans and the increase in water vapour would enhance global warming, increasing the temperature still further, leading to further vaporisation of water from the oceans and so on.

The results of global warming would include melting of polar ice caps and flooding of low-lying countries.

In 1992 the Panel published a second report from 170 scientists in 47 countries in which they admitted that further computer modelling had not confirmed the predictions of the 1990 report. The Panel gave a revised estimate of global warming; between 1 °C and 4.5 °C . It stated that the increase in temperature over the past 100 years fell within the normal variation in the climate and might even be due to causes other than carbon dioxide. The computer modelling now took into account the interaction between the atmosphere and the ocean but did not include clouds. A new factor was introduced: sulphate dust from industrial emissions may have a cooling effect. The model could not account for the disappearance of 7 billion tonnes of the carbon dioxide that enters the atmosphere each year [§ 2.9.1]. The 1992 report stated that the estimates of global warming over the past 100 years due to greenhouse gases made in the 1990 report were too high. The report stated that better temperature measurements were needed as well as better computer simulations.

In 1992 the Panel issued a second report in which they revised their estimate of global warming and asked for more research work to be done.

Methods of disposing of excess carbon dioxide have been suggested.

Most research continues to be on computer simulations. Some groups with a more practical bent have suggested methods of disposing of carbon dioxide. One suggestion is to collect the carbon dioxide produced when fossil fuels are burned, liquefy it and pump it into the oceans. At a depth of 400 m, the pressure and temperature are such as to keep carbon dioxide in liquid form. There is, however, a danger that a sudden vaporisation of carbon dioxide from the depths of the sea would cause an enormous tidal wave that would swamp everything in its path. Another suggestion is to store carbon dioxide on land as giant snowballs of solid carbon dioxide. A disadvantage is that to provide the energy needed to keep carbon dioxide solid an enormous additional amount of energy would need to be generated.

The Earth Summit of 1992 and the Berlin Mandate of 1992 agreed to reduce emissions of greenhouse gases.

At the 1992 **Earth Summit**, 2000 nations agreed to stabilise carbon dioxide emissions at the 1990 levels. This would not be enough if the worst warnings of global warning were true. At a conference of industrial nations in Berlin in 1995, the **Berlin Mandate** agreed to continue to reduce emissions of all greenhouse gases after the year 2000.

1. You often hear people talking about the greenhouse effect as if it were a pollution phenomenon. Explain why the greenhouse effect is in fact beneficial to the human race.

2. It has been suggested that a number of 'multiplier effects' could make an enhancement of the greenhouse effect worse than expected. Explain how the following events would add to global warming.

(*a*) If the polar ice caps began to melt, the land beneath them would be exposed.

(*b*) A rise in temperature would lead to evaporation of water from the oceans.

(*c*) A rise in temperature would decrease the solubility of carbon dioxide.

(*d*) A rise in temperature might add to volcanic activity.

3. (*a*) How does deforestation add to concern over the greenhouse effect?

(*b*) Which countries are cutting down trees to free land for agricultural use?

(*c*) Suppose that experts were to state that the present rate of carbon dioxide emission would certainly cause a serious rise in the Earth's temperature with catastrophic results. If it were up to you to persuade countries to stop deforestation and sacrifice their short-term ends for the long-term future of the planet, what would you say? What resistance do you think you might meet?

4. (*a*) Describe two ways in which carbon dioxide is emitted into the atmosphere.

(*b*) Explain why increased levels of carbon dioxide could contribute to global warming.

(*c*) Why is it difficult to know for certain whether the Earth's temperature is rising or not?

(*d*) Give equations for two chemical reactions by which carbon dioxide is removed from the atmosphere.

5. (*a*) The monthly average carbon dioxide levels in the Northern Hemisphere recorded in Figure 2.9C show that the value falls in the summer and rises in the winter. Explain this variation.

(*b*) Suggest two regions on Earth where you would expect the annual variation to be smaller than shown in Figure 2.9C.

(*c*) There is an upward trend in the long term. Explain why this is happening.

(*d*) Explain why many people are concerned about this upward trend in the level of carbon dioxide.

2.10 THE OZONE LAYER

There is a layer of ozone in the stratosphere.

The sunlight that falls on the stratosphere [see Figure 2.3A] contains much more ultraviolet (UV) light than the radiation which reaches the surface of the Earth. Ultraviolet light has enough energy to bring about photochemical reactions that convert dioxygen, O_2 into ozone, O_3. As plant life evolved on Earth, photosynthesis increased the concentration of oxygen in the atmosphere. In consequence, the concentration of ozone in the stratosphere increased until equilibrium was reached between the rate of formation of ozone from oxygen and the rate of decomposition of ozone. The maximum concentration of ozone, about 10 ppm, occurs 25–50 km from the surface of the Earth.

The first step in the formation of ozone is the dissociation of O_2 molecules by UV light to form O·

Some of the reactions involving dioxygen, O_2, and ozone (trioxygen, O_3) are:

(*a*) Dioxygen is dissociated by solar UV rays:

$$O_2 + h\nu(UV) \longrightarrow O\cdot + O\cdot$$

(*b*) Some of the oxygen atoms formed combine with dioxygen molecules to form ozone:

Then O_2 reacts with O· to form O_3.

$$O\cdot + O_2 + M \longrightarrow O_3 + M^*$$

(*c*) Ozone absorbs UV light and dissociates:

$$O_3 + h\nu(UV) \longrightarrow O_2 + O\cdot$$

(*d*) Ozone reacts with oxygen atoms to form dioxygen

$$O_3 + O\cdot \longrightarrow 2O_2$$

Reactions (*b*), (*c*) and (*d*) are exothermic. The heat produced in the stratosphere explains why a temperature inversion takes place at the tropopause [see Figure 2.3A].

There is a natural balance which keeps the ozone layer at a constant thickness.

The balance between the rates of formation and destruction result in a steady state concentration of ozone in the stratosphere. In the natural course of events, if the thickness of the ozone layer decreases, UV light can penetrate further through the stratosphere to reach dioxygen molecules, and the rates of photolysis of oxygen and formation of ozone increase. The result is that the ozone layer becomes thicker again and less UV light can penetrate to reach oxygen molecules so the rates of photolysis of oxygen and formation of ozone decrease. At a certain concentration of ozone, the rate of formation of ozone is equal to the rate of dissociation of ozone to form oxygen: equilibrium has been reached.

The ozone layer prevents excessive UV light from reaching the troposphere.

The ozone in the stratosphere is present as a thin **ozone layer**. This would be only 3 mm thick if measured at atmospheric pressure and temperature at ground level! Ozone absorbs ultraviolet radiation strongly at wavelengths of 220–330 nm. It is effective in filtering out dangerous UV-B radiation (290–320 nm) . UV-A radiation at 320–400 nm is less harmful, and UV-C radiation at wavelengths less than 290 nm does not penetrate to the troposphere. There is a strong connection between UV radiation and the incidence of both non-fatal and fatal skin cancer in humans. The ozone layer protects us from this damaging radiation.

Ozone is destroyed by chemical reactions with a number of substances that occur naturally in the stratosphere, e.g. nitrogen oxides and methane. Free radicals are formed by photodissociation, e.g.

$$O_2 + hv \longrightarrow 2O\cdot$$

$$N_2O + hv \longrightarrow N_2 + O\cdot$$

$$CH_4 + hv \longrightarrow \cdot CH_3 + H\cdot$$

Free radicals that react with ozone include hydrogen atoms:

Ozone is destroyed by reaction with H·, NO·, O· and other radicals.

$$H\cdot + O_3 \longrightarrow \cdot OH + O_2$$

$$\cdot OH + O\cdot \longrightarrow O_2 + H\cdot$$

and oxides of nitrogen:

$$NO\cdot + O_3 \longrightarrow NO_2\cdot + O_2$$

$$NO_2\cdot + O \longrightarrow NO\cdot + O_2$$

Some species are present at very low concentrations; yet they may be important if their rates of reaction are high. Compare the rate constants of these two reactions:

$$O\cdot + O_3 \longrightarrow 2O_2; \ k = 5 \times 10^5 \, \text{dm}^3 \, \text{mol}^{-1} \, \text{s}^{-1}$$

$$O\cdot + NO_2 \longrightarrow NO\cdot + O_2; \ k = 5 \times 10^9 \, \text{dm}^3 \, \text{mol}^{-1} \, \text{s}^{-1}$$

The high value of the rate constant for the second of these reactions illustrates why oxides of nitrogen are very important in the stratosphere. They arise from natural sources; in their absence the ozone layer would be twice as thick as it is. They are also generated by the combustion of fossil fuels. The oxide NO does not reach the stratosphere because it is too reactive. The oxide N_2O is sufficiently unreactive to reach the stratosphere where NO is generated from it:

$$\text{Source: } O\cdot + N_2O \longrightarrow 2NO\cdot$$

$$\text{Sink: } \cdot OH + NO_2 \longrightarrow HNO_3 \longrightarrow \text{Acid rain} \ [\S 3.6]$$

Man-made substances are now attacking the ozone layer.

Increasing quantities of man-made substances are now attacking the ozone layer and reducing its thickness. The effect of atmospheric pollutants on the ozone layer is discussed in § 3.11.

CHECKPOINT 2.10

1. (*a*) Why is the ozone layer referred to as a protective 'blanket' round the Earth?

(*b*) What natural factors limit the thickness of the ozone layer?

2. (*a*) How is ozone formed from oxygen in the stratosphere?

(*b*) Write equations for two reactions that destroy ozone.

3. What is (*a*) the source and, (*b*) the sink of NO in the stratosphere?

QUESTIONS ON CHAPTER 2

1. Explain why there is

(*a*) a temperature minimum at the boundary of the troposphere and the stratosphere

(*b*) a temperature maximum at the boundary of the stratosphere and mesosphere.

2. Describe the role of micro-organisms in the nitrogen cycle.

3. (*a*) Describe the main features of the carbon cycle.

(*b*) How is it related to the oxygen cycle?

(*c*) Say what process is represented by the reaction (in which (CH_2O) represents organic material):

$$2(CH_2O) \longrightarrow CO_2(g) + CH_4(g)$$

4. (*a*) What is the function of a 'third body' in an atmospheric chemical reaction?

(*b*) What is meant by the symbols · and * in representing chemically active species in the atmosphere?

(*c*) Which of the following species could most readily be transformed into a 'normal' species in isolation from other reactants?

$O·$ $HO·*$ $NO_2·*$ $H_3C·$ $N*$

5. State two factors that make the stratosphere important as a region where atmospheric trace contaminants are converted into other, less reactive substances.

6. What two chemical species are most often responsible for removal of OH · radicals from an unpolluted troposphere?

7. Anaerobic fermentation of organic matter (CH_2O) in water yields $15.0 \, dm^3$ of CH_4 at rtp. What volume of O_2 would be consumed by aerobic respiration of the same quantity of (CH_2O)?

3

AIR POLLUTION

Atmospheric pollutants are a serious threat to health.

Waste matter is released into the atmosphere from a variety of sources. Many types of waste cause damage to the environment and to living organisms. We call these substances **pollutants**. The pollutants present in the atmosphere are particulate solids, droplets of liquids and gases. An average person inhales between 10 000 and 20 000 dm^3 of air in 24 hours. The rate increases with exercise; a jogger may inhale 3000 dm^3 in one hour. The presence of pollutants in the atmosphere is therefore a very serious matter; there is no escaping from inhaling air.

They are primary pollutants . . . . . . and secondary pollutants.

It is convenient to classify pollutants as **primary pollutants** which are emitted directly into the atmosphere and **secondary pollutants** which are formed in the atmosphere from primary pollutants.

3.1 PARTICLES

Particles include . . . . . . solid particles of carbon, dust, coal, etc. . . . . . . liquid particles of mist, fog, etc. . . . . . . biological particles, e.g. viruses, bacteria.

The troposphere contains a large mass of particles, both solid and liquid. Solid particles of 0.001–10 μm diameter are suspended in the air near sources of pollution, e.g. fossil fuel power plants and busy roads. They include carbon, soil dust, foundry dust and pulverised coal. Liquid particles include mist, including raindrops, fog and sulphuric acid mist. Particles of biological origin include viruses, bacteria, bacterial spores, fungal spores and pollen. As particles diffuse through the air, smaller particles may coagulate to form larger particles and precipitate from the atmosphere. Some sources of particulate pollution are listed.

Particulate pollutants are formed by: mechanical processes, e.g. grinding . . .

1. Mechanical processes which give rise to particles include coal grinding, the formation of spray in cooling towers, dirt blowing off dry soil, road-making, quarries and rock crushers.

. . . chemical processes, e.g. power plants, cement kilns and the combustion of fuels.

2. Chemical processes that produce particles are combustion processes, such as fossil fuel power plants, incinerators, furnaces, fireplaces, cement kilns, internal combustion engines, forest fires and active volcanoes. Particles formed from combustion sources tend to be < 1 μm in diameter and are readily carried into the lungs. They can become coated with more hazardous components such as toxic heavy metals .

3. Metal oxides are one type of inorganic particle in the atmosphere. They are formed whenever fuels containing metals are burned, e.g. iron(III) oxide particles are formed when coal containing iron(II) sulphide is burned.

4. Aerosol mists are formed when atmospheric sulphur dioxide is oxidised to sulphuric acid, which, being hygroscopic, combines with atmospheric water vapour to form small liquid droplets. Basic pollutants in the air, e.g. ammonia and calcium oxide, can react to form salts – ammonium sulphate and calcium sulphate – and when water is lost by evaporation a solid aerosol is formed.

Mists are formed by reactions of atmospheric sulphur dioxide.

5. Of **organic particles,** a large proportion are produced by internal combustion engines. They include alkanes, arenes, carboxylic acids, benzoic acids, aromatic aldehydes and carcinogenic polycyclic aromatic hydrocarbons, such as:

Carcinogens come from vehicle engines.

Benzopyrene   Chrysene   Benzofluoranthene

6. Carbon as soot, carbon black, coke and graphite originates from vehicle exhausts, heating furnaces, incinerators, power plants and steel foundries. Carbon is one of the most visible air pollutants. It has good adsorbent properties and is a carrier of other pollutants. The surfaces of carbon particles catalyse atmospheric reactions , such as the conversion of sulphur dioxide into sulphuric acid.

Carbon comes from vehicle engines, power plants and incinerators.

7. Fly ash: In the combustion of fossil fuels in a furnace, much of the inorganic matter is converted into a fused ash which does not contribute to pollution. Smaller particles of fly ash leave the furnace with the exhaust gases and are filtered out by an efficient stack system. Some fly ash escapes, however, and enters the atmosphere; this part of the fly ash consists of small particles which do the most damage to human health, plants and visibility. Components of fly ash include carbon and oxides of metals.

Fly ash comes from the combustion of fossil fuels in furnaces.

8. Asbestos is a silicate of fibrous structure. It is strong, flexible and non-flammable. These properties lead to uses as a structural material, brake linings, insulation and pipe manufacture. In 1980, the consumption of asbestos was high, yet by 1990 most uses of asbestos had been phased out. The severity of asbestos as an air pollutant is that when inhaled it causes asbestosis (a pneumonia condition), tumours of the tissue lining the chest cavity and lung cancer.

Asbestos fibres damage the lungs.

9. Toxic metals found as particulate matter in polluted atmospheres are 'heavy metals' and beryllium. Lead is the toxic metal which reaches the most worrying levels in urban atmospheres [§ 3.2], and mercury ranks second. Others are beryllium, cadmium, chromium, vanadium, nickel and arsenic. In time the particles precipitate to contaminate the ground and are leached into waterways and groundwater.

Mercury: Some of the mercury in the atmosphere is adsorbed on particles. Much of the mercury entering the atmosphere does so as mercury vapour from the combustion of coal and from volcanoes. [See also mercury as a water pollutant § 6.9]

Toxic metals in particulate pollutants include lead, beryllium, mercury cadmium, chromium, vanadium, nickel and arsenic.

Beryllium: Beryllium was used in phosphors in fluorescent lamps until, during the 1940s and 1950s, it was found that beryllium and its compounds are toxic. More recently, a number of high-technology applications have been found for beryllium in electrical equipment, electronic instruments and nuclear reactors and consumption of beryllium may increase in the future.

3.1.1   EFFECTS OF PARTICLES

An obvious effect of atmospheric particles is that they reduce visibility by scattering light. They reflect radiation from space and prevent it from reaching the Earth's surface, thus exerting a cooling effect on the Earth. They also provide surfaces on

Particulate pollutants reduce visibility . . .

*... and catalyse
atmospheric reactions.*

which heterogeneous atmospheric chemical reactions can occur – thus exerting an effect on air pollution – and nuclei for the condensation of atmospheric water vapour, thus influencing the weather.

Atmospheric particles inhaled into the respiratory tract may damage health. Relatively large particles are likely to be retained in the nasal cavity and the pharynx, whereas very small particles are likely to reach the lungs. The respiratory system has a mechanism for expelling inhaled particles: cilia carry the particles on a bed of mucus to the entrance to the gastrointestinal tract where they are swallowed. When particulate matter enters the lungs it is not only the respiratory system that may be damaged because the blood transports particles to other organs also. It has been observed that an increase in the daily mortality rate follows a severe episode of atmospheric pollution. This may not be due to the high levels of particulate pollution alone because high levels of sulphur dioxide and other pollutants occur at the same time.

*When large particles are
inhaled, they are retained
in the nasal cavity and later
swallowed. Small particles
penetrate to the lungs.*

3.1.2 METHODS OF REMOVING PARTICLES

SEDIMENTATION

*Particles can be removed
from a stream of gas by
sedimentation ...*

The simplest method of removing particles from a gas stream is sedimentation, in which particles settle under the influence of gravity. Gravitational settling chambers take up a lot of space and are not very effective with small particles.

FILTRATION

Fabric filters allow the passage of gas but retain particles. They are used to collect dust in structures called baghouses. From time to time the filter is shaken to remove particles.

... by filtration ...

SCRUBBERS

... by scrubbers ...

A stream of the 'scrubbing' liquid injected at right angles to the incoming gas removes particles from the gas stream [see Figure 3.1A].

FIGURE 3.1A
A Wet Scrubber

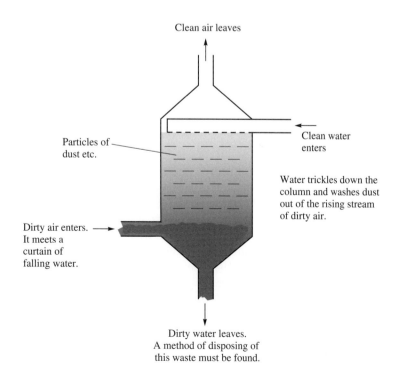

Clean air leaves

Particles of
dust etc.

Clean water
enters

Water trickles down the
column and washes dust
out of the rising stream
of dirty air.

Dirty air enters.
It meets a
curtain of
falling water.

Dirty water leaves.
A method of disposing of
this waste must be found.

Figure 3.1B
An Electrostatic
Precipitator

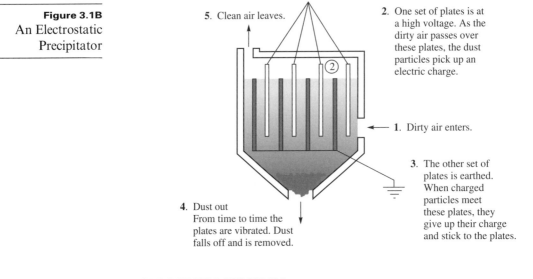

Figure 3.1B
An Electrostatic
Precipitator

5. Clean air leaves.

2. One set of plates is at a high voltage. As the dirty air passes over these plates, the dust particles pick up an electric charge.

1. Dirty air enters.

3. The other set of plates is earthed. When charged particles meet these plates, they give up their charge and stick to the plates.

4. Dust out
From time to time the plates are vibrated. Dust falls off and is removed.

ELECTROSTATIC REMOVAL

... by electrostatic precipitation.

Aerosol particles may acquire electrical charge when passed through a high-voltage electric field. The charge causes the particles to be attracted to an oppositely charged surface from which they may later be removed [see Figure 3.1B].

CHECKPOINT 3.1

1. Why are smaller particles more effective than larger particles per unit mass in catalysing atmospheric chemical reactions?

2. Why are smaller particles more dangerous than larger particles per unit mass when they are inhaled?

3. List four industries which give rise to particulate pollutants.

4. List four methods used in industry for the removal of pollutants.

3.2 LEAD

Lead enters the air from lead tetraethyl, which is added to petrol as an anti-knock.

Tetraethyllead, $Pb(C_2H_5)_4$ has been added to petrol since 1945 to prevent knocking – a metallic rattle caused by uneven combustion of petrol vapour. To prevent the build-up of lead deposits in the engine, dichloroethane and dibromoethane, $C_2H_4Cl_2$ and $C_2H_4Br_2$, are added. They react with lead to form volatile lead halides which are removed from the engine in the exhaust gases. In the early 1970s about 200 000 tonnes of lead entered the atmosphere each year from the USA alone.

The level of lead in Greenland ice rose sharply after the introduction of leaded petrol.

It is difficult to find any place on Earth with natural levels of lead. To give a comparison of atmospheric lead concentrations through the years, scientists have analysed ice which solidified in Greenland long ago and has never melted. The depth of a layer of ice below the surface is a measure of its age. Long vertical columns of ice have been drilled from deep deposits of ice in Greenland and analysed for lead. Ice deposited from snow that fell before 1750 contained less than 0.001 ppb of lead. From 1750 to 1940 the concentration of lead increased steadily to 0.08 ppb. In 1940 tetraethyllead was first introduced into petrol on a large scale. Snow that fell in 1950 contained 0.20 ppb of lead – an increase of 250 per cent. The level continued to rise until the introduction of unleaded petrol in 1980–90 reduced this source of pollution.

The lead compounds in vehicle exhausts are present as fine particles $< 2\,\mu m$ in diameter, which can be dispersed over a wide area. When they are inhaled into the

The results of inhaling lead compounds include anaemia, headaches and depression . . .

. . . followed by damage to kidneys, reproductive system, liver and brain.

lungs about 40% of the lead is absorbed to enter the bloodstream. There it has a residence time of one month, before it enters bones where it has a residence time of 40–90 years. Lead bonds strongly to haemoglobin, DNA, RNA and many enzymes and can therefore disrupt many metabolic pathways. Mild lead poisoning causes anaemia, headaches, depression and fatigue. More severe effects are damage to kidneys, reproductive system, liver and brain.

At its worst, the distribution of lead compounds from motor vehicles increased atmospheric levels of lead by a factor of 20 (more in urban areas). It pollutes the land as well as the air: soil and crops are contaminated when the particles of lead compounds fall to Earth, and the lead content of food rises. Studies in the 1970s found that roadside dust contained 100–3000 ppm of lead. Children are particularly at risk from lead poisoning as they play in the dirt and transfer dirt from their hands to their mouths. In the 1970s, Harvard Medical School in the USA identified children who had been exposed to high atmospheric concentrations of lead by the lead content of their teeth. They found that these children performed less well in intelligence tests and their ability to concentrate and that their behaviour suffered. In 1980, researchers at Reading University and Birmingham University in the UK put forward the view that lead from vehicle exhausts was the chief source of lead poisoning. They suggested that aggressive behaviour in urban children may be caused by the effects of lead on the brain.

Lead from the atmosphere pollutes the land.

Vehicle exhausts are the chief source of lead pollution.

Lead-free petrol is now used in vehicles manufactured after 1990 in the UK and after 1991 in Hong Kong.

The Campaign for Lead-Free Air, CLEAR, saw the fruits of its 20-year fight in 1990. Since 1990 all new vehicles in the UK are manufactured to run on lead-free petrol. Leaded petrol is still available for older vehicles. The USA and Germany were several years ahead, France was slower to change. In Hong Kong unleaded petrol became available in 1991. Another advantage of unleaded petrol is that cars can be fitted with catalytic converters to reduce the amounts of other pollutants in exhaust gases [see §3.9].

3.3 CARBON MONOXIDE

Carbon monoxide is one of the inorganic pollutants in the air. Others are sulphur dioxide, nitrogen oxides, ammonia, hydrogen sulphide, chlorine, hydrogen chloride and hydrogen fluoride.

Carbon monoxide is poisonous.

Carbon monoxide is toxic. By forming a complex with iron(II) in haemoglobin (*ALC*, §24.13.8), it prevents haemoglobin from combining with oxygen. The normal atmospheric concentration is 0.1 ppm, and the residence time is 36–110 days. If the concentration of carbon monoxide is 600–700 ppm, inhalation for 1 hour gives barely noticeable symptoms. If the concentration is 4000 ppm, inhalation causes death inside 1 hour. Exposure to a concentration of 1000 ppm for 4 hours is likely to cause death. Being colourless and odourless, carbon monoxide gives no warning of its presence.

It is formed naturally by the oxidation of methane.

Some of the carbon monoxide in the atmosphere is formed by the oxidation of methane, which is formed naturally by the degradation of plant material. Carbon monoxide is emitted from the internal combustion engine; therefore the highest levels of carbon monoxide occur in urban areas during rush hours, when levels may reach 50–100 ppm. The control of emissions from the internal combustion engine is the best hope of reducing carbon monoxide levels. A leaner air–fuel mixture may be employed – one with a high ratio of air/fuel [see §3.9]. The fitting of catalytic converters cuts the emission of carbon monoxide [see §3.9].

Carbon monoxide is emitted from vehicle engines so that concentrations in urban areas in rush hours are high.

Carbon monoxide is removed from the atmosphere by reaction with hydroxyl radicals.

Carbon monoxide is removed from the atmosphere by reaction with · OH radicals ...

$$CO + HO \cdot \longrightarrow CO_2 + H \cdot$$

followed by

$$O_2 + H \cdot + M \longrightarrow HOO \cdot + M$$

and the regeneration of HO · by reactions such as

$$2HOO \cdot \longrightarrow H_2O_2 + O_2$$

$$H_2O_2 + h\nu \longrightarrow 2HO \cdot$$

The oxidation occurs slowly because the energy of activation is high. If there were no other mechanism for the destruction of carbon monoxide it would occur as a toxic blanket round the Earth. The crucial oxidation occurs in the soil. Several soil micro-organisms have enzymes which catalyse the oxidation and remove carbon monoxide from the atmosphere; soil is a **sink** for carbon monoxide.

... and, more importantly, by micro-organisms in the soil.

CHECKPOINT 3.3

1. (*a*) Why is carbon monoxide a dangerous pollutant?

(*b*) How is it removed from air by natural processes?

(*c*) What relatively recent source of carbon monoxide has led to increased atmospheric levels of carbon monoxide?

2. (*a*) What is the advantage of adding lead compounds to petrol?

(*b*) What is the disadvantage?

(*c*) What happened when the UK Government decided that the disadvantages outweighed the advantages?

3.4 SULPHUR DIOXIDE

Sulphur dioxide enters the atmosphere from the combustion of fossil fuels and from volcanoes ...
... biological decay of organic matter and from the reduction of sulphates.

Sulphur dioxide enters the atmosphere from the combustion of fossil fuels, from volcanoes, from biological decay of organic matter and the reduction of sulphates. Any hydrogen sulphide that enters the atmosphere is rapidly oxidised to sulphur dioxide. Coal contains sulphur as iron sulphides and as organic compounds. To keep the emission of sulphur dioxide down to acceptable levels, sulphur must be removed from coal.

Sulphur dioxide reacts to form sulphuric acid and sulphates. The products are present as aerosol particles and as solid particles and reduce visibility. Sulphur dioxide reacts in the atmosphere in a number of ways:

Sulphur dioxide and oxides react with water droplets to form acid rain.

- chemical and photochemical reactions in the presence of hydrocarbons and oxides of nitrogen [§ 3.8]

- chemical reactions inside water droplets and on the surface of solid particles in the atmosphere which lead to the formation of acid rain

Sulphur dioxide reacts with hydrocarbons, oxides of nitrogen and UV light to form photochemical smog ...

Different processes predominate under different conditions. Light of wavelength > 218 nm is not sufficiently energetic to bring about photodissociation of sulphur dioxide so direct photochemical reactions in the troposphere do not occur. The oxidation of sulphur dioxide in otherwise unpolluted air is a slow process. The presence of other pollutants – hydrocarbons and oxides of nitrogen – greatly increases the rate of oxidation of atmospheric sulphur dioxide. For the formation of photochemical smog, hydrocarbons, oxides of nitrogen and UV light must be present. Ozone is present in photochemical smog; it oxidises sulphur dioxide slowly in the gas phase but more rapidly in water droplets. Soot particles are effective in catalysing the oxidation of sulphur dioxide to sulphates.

... containing sulphuric acid and sulphates. Sulphur dioxide damages the respiratory system.

Sulphur dioxide damages the respiratory system and causes an increase in asthma attacks. People differ greatly in their susceptibility to sulphur dioxide.

3.5 OXIDES OF NITROGEN

3.5.1 DINITROGEN OXIDE

Dinitrogen oxide, N_2O, is produced naturally and passes to the stratosphere.

Dinitrogen oxide, N_2O, is produced by microbiological processes. It is relatively unreactive and passes from the troposphere into the stratosphere where it is photolysed into nitrogen and nitrogen monoxide.

3.5.2 NITROGEN MONOXIDE AND NITROGEN DIOXIDE

Nitrogen monoxide, NO, and nitrogen dioxide, NO_2, are represented as NO_x.

Nitrogen monoxide, NO, and nitrogen dioxide, NO_2, together represented as NO_x, enter the atmosphere from natural sources such as lightning discharges and biological processes and from pollutant sources. The combustion of fossil fuels gives rise to most of the NO_x from man-made sources; 100 million tonnes of NO_x a year come from this source. The quantity of NO_x that enters the atmosphere from natural sources is many times this quantity.

They enter the atmosphere from lightning discharges . . . . . . from biological processes . . . . . . and from the combustion of fossil fuels.

Much of the NO_x entering the atmosphere is from the internal combustion engine. At the high temperatures inside the cylinders nitrogen and oxygen combine to form nitrogen monoxide.

$$N_2 + O_2 \rightleftharpoons 2NO$$

The reaction is endothermic, and both the rate of reaction and the equilibrium constant increase steeply as the temperature increases.

Hydrocarbons are also present in the combustion mixture. They react with oxygen atoms to form hydroxyl radicals which take part in the formation of nitrogen monoxide.

$$RH + O\cdot \longrightarrow R\cdot + HO\cdot$$

$$N\cdot + HO\cdot \longrightarrow NO\cdot + H\cdot$$

The formation of NO_x in vehicle engines is higher when the air/fuel ratio exceeds the stoichiometric ratio.

The formation of nitrogen monoxide is higher when the air/fuel ratio exceeds the stoichiometric ratio for the combustion of the fuel. In § 3.9 the relationships between pollutants from the internal combustion engine are discussed.

FIGURE 3.5A
Reactions of NO_x

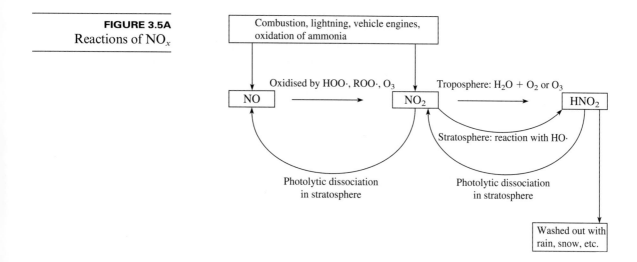

3.5.3 REACTIONS OF NO$_x$ IN THE ATMOSPHERE

NO$_x$ are converted into acid rain ...
... and into PAN in photochemical smog.

Reactions in the atmosphere convert NO$_x$ into nitric acid, inorganic nitrates, organic nitrates and peroxyacetylnitrates, PAN. They play a serious part in the formation of photochemical smog [see § 3.8].

3.6 ACID RAIN

Rain is naturally slightly acidic because it reacts with carbon dioxide in the air to form carbonic acid. Natural rainwater has a pH of about 5.6. In central Europe, rainwater is much more acidic, with a pH of about 4.1, and on the fringes of Europe, e.g. Ireland and Portugal, rainwater has a pH of about 4.9. Rain from individual storms can have a pH below 3, and water droplets in fogs can be even more acidic.

All rainwater is slightly acidic. Acid rain has a pH below 5.0.

Acid rain is now thought to be the cause of the extensive damage to Europe's trees and to the death of fish in the lakes of Canada, Norway, Sweden, Wales, Scotland and other countries. Europe shows the worst signs of damage by acid rain, but acid rain is becoming a global phenomenon.

FIGURE 3.6A
The Effects of Acid Rain on Trees

Lakes in many countries of Europe and North America have become so acidic that their stocks of fish have died.

Lakes in Scandinavia, Scotland, Canada and the USA have become much more acidic. Thousands of lakes which once stocked fish are now dead. The death of fish is usually attributed to poisoning by aluminium. Sulphate ions in acid rain can combine with aluminium in complex compounds to form soluble aluminium sulphate, which washes into streams. There it interferes with the operation of fish gills, so that they become clogged with mucus. The fish die from lack of oxygen.

FIGURE 3.6B
A Statue Attacked by
Acid Rain

Many of Europe's trees have lost much of their foliage. In some countries large areas of forest have died. The cause is thought to be acid rain.

Forests have declined. In the mid-1970s Alpine forests started to lose their fir trees. Then in Germany, including the famous Black Forest, spruce began to thin and their needles turned brown. After 1984, the decline stabilised in Germany. The Netherlands, Czechoslovakia, Switzerland and Britain recorded that 20–30% of their trees were severely defoliated. Acidic rain-water draining from soils washes out nutrients and liberates aluminium ions [see § 7.9] which the roots of trees may take up. Without essential nutrients, e.g. calcium and magnesium, trees starve to death.

Building materials which are basic . . . and iron and steel . . . are attacked by acid rain.

Building stone may be limestone or marble (both calcium carbonate) or a sandstone in which quartz grains are held together by a coating of calcium carbonate or iron oxide. These materials (apart from quartz) are attacked by acid rain to form products which are water-soluble or in the form of poorly adherent surface crusts. Metallic structures, e.g. bridges, ships and motor vehicles, are also attacked by acid rain.

FIGURE 3.6C
The Formation of
Acid Rain

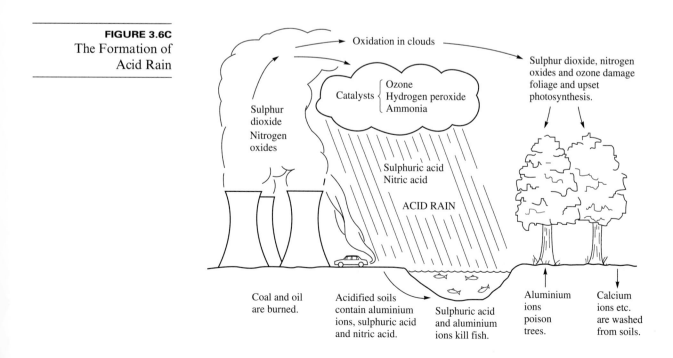

Sulphur dioxide is one of the causes of acid rain. It is formed in the combustion of fossil fuels. In the atmosphere, sulphur dioxide reacts with oxygen and water vapour to form sulphuric acid.

One of the chief culprits in the formation of acid rain is sulphur dioxide. Natural sources, such as volcanoes, sea spray, rotting vegetation and plankton, send sulphur dioxide into the atmosphere. Half the sulphur dioxide in the atmosphere, however, comes from the combustion of fossil fuels. Over Europe the proportion of sulphur dioxide in the air that comes from fuels is 85%. When sulphur dioxide reaches the atmosphere, it reacts with moisture and oxygen to form sulphuric acid. Not all the sulphur dioxide is converted into sulphuric acid: in dry air, sulphur dioxide can travel hundreds of kilometres with little conversion to acid, and can descend to ground level unconverted. When sulphur dioxide is incorporated into clouds, conversion into sulphuric acid takes place within two hours. In heavily polluted city air, tiny particles of metals catalyse the reactions. Ozone and hydrogen peroxide also seem to assist them, as does ammonia, which is given off in large quantities from slurry tanks.

FIGURE 3.6D
Sources of
Sulphur Dioxide

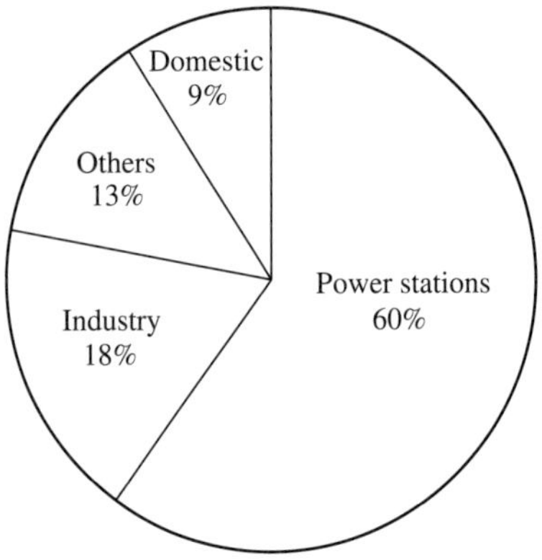

The 'Thirty Percent Club' of nations have agreed to reduce their emission of sulphur dioxide.

Acid rain has become an important political issue, souring relationships between countries which export pollution and those which receive it; for example, between Britain and Norway and between the USA and Canada. In 1984 the UK joined the 'Thirty Percent Club', a group of nations committed to reducing sulphur dioxide emission by 30% by the year 2000. Figure 3.6D shows that the biggest source of sulphur dioxide is power stations. In 1984, the Central Electricity Generating Board (CEGB) estimated a cost of £120 million for treating the exhaust gases from each existing power station and an increase in running costs of 10–15% which would increase the price of electricity by 5%.

Several measures can be taken to reduce this pollution.

LOW SULPHUR FUELS

Some of the sulphur in oil and coal can be removed before the fuels are burnt.

Oil power stations emit on average about 10% more sulphur dioxide than coal power stations. Modern oil refineries can, however, produce low-sulphur oil. The coal used in British power stations contains an average of 1.6% sulphur. About half the sulphur is combined as 'iron pyrites', FeS_2. Of this, 80% can be removed by grinding the coal and using various separation techniques. Removal of organically bound sulphur is more difficult. If the coal is converted into a gaseous fuel by one of the modern methods of *coal gasification*, sulphur can be removed as the gas is formed. Coal gasification has another advantage in that it may supplement our diminishing reserves of natural gas. Of the sulphur in bituminous coal, 25% can be removed by crushing and cleaning. A number of countries in Eastern Europe use lignite which cannot be cleaned by present technology.

NEW BURNERS

Pulverised fluidised bed combustion (PFBC) is a technique for removing sulphur from coal as it burns. PFBC can be built into new power stations.

Pulverised fluidised bed combustion (PFBC) [see Figure 3.6E] offers a new, more efficient way of burning coal in a bed of limestone, which removes sulphur as the coal burns. PFBC can cut the emission of nitrogen oxides as well as sulphur dioxide. Pilot tests suggest that such burners could remove up to 80% of sulphur dioxide. These systems may soon be widely used in pollution-conscious countries.

PFBC has the advantage over other systems [for FGD, see below] that PFBC is cheaper and more energy-efficient and creates fewer problems in the disposal of waste sludges. British Coal is developing the technology in its Grimethorpe research plant in Yorkshire, but PFBC will not be ready in time to achieve the needed reduction in emission by the year 2000, and an alternative method (flue-gas desulphurisation; see below) is already available.

FIGURE 3.6E
Pulverised Fluidised Bed
Combustion (PFBC)

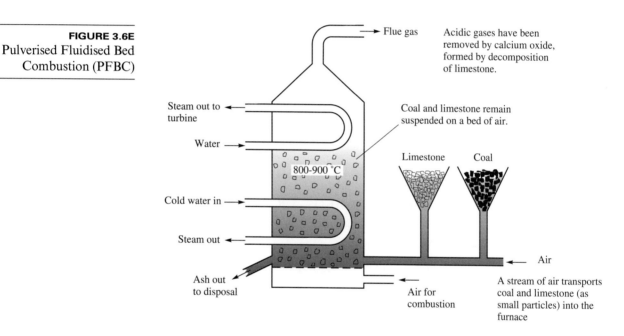

REMOVAL OF SULPHUR FROM EXHAUST GASES

A large power plant emits about 10^8 m^3 of gases daily — roughly equal to the volume of air in 100 000 houses. Tall chimneys take the exhaust gases, including sulphur dioxide and other pollutants, high up into the air before they are discharged. The work force and the local community are protected. At one time, the managers of power stations and factories thought that was the end of the problem. Now that we know acidic pollutants are to blame for the acid rain that falls perhaps thousands of kilometres away, we realise that tall chimneys do not solve the problem; they merely transfer it to another region.

Flue gas desulphurisation (FGD) is a technique for removing sulphur dioxide from the exhaust gases in the chimney stacks of power stations.

Sulphur dioxide can be removed from the exhaust gases before they leave the chimney stack of the power station. Systems that carry out such **flue gas desulphurisation** (FGD) have been developed and are increasingly used in power stations. The exhaust gases are washed by an alkaline solution, which converts the sulphur dioxide into a waste sludge. The systems can remove up to 95% of the sulphur in the flue gases. [see Figure 3.6F].

FIGURE 3.6F
Flue Gas
Desulphurisation (FGD)

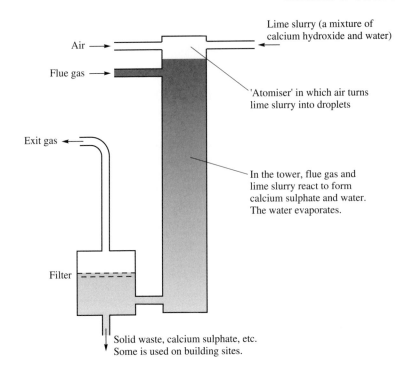

The concentration of sulphur dioxide in flue gases from power stations is about 0.3%. Some of the alkaline solutions and suspensions used to remove it are as follows:

(1) A slurry of limestone and lime is used to 'scrub' the flue gases.

(*a*) $CaCO_3 + SO_2(g) \longrightarrow CaSO_3 + CO_2(g)$

(*b*) $CaO + SO_2 \longrightarrow CaSO_3$

(*c*) $2CaSO_3 + O_2 + 4H_2O \longrightarrow 2CaSO_4 \cdot 2H_2O$

Reactions (*a*), (*b*) and (*c*) produce an impure sludge which is dumped.

(2) A slurry of magnesium oxide is used as a 'scrubber'.

(*d*) $MgO + SO_2 \xrightarrow{\;H_2O\;} MgSO_3 \xrightarrow{\;heat\;} MgO + SO_2$

The magnesium sulphite produced is heated to give magnesium oxide, which is recycled, and sulphur dioxide at a concentration high enough to allow it to be used in the manufacture of sulphuric acid.

(3) A solution of sodium sulphite can be used for 'scrubbing'.

(*e*) $Na_2SO_3 + H_2O + SO_2 \longrightarrow 2NaHSO_3$

The sodium hydrogensulphite produced can be heated to give sodium sulphite for recycling and sulphur dioxide for sale to sulphuric acid manufacturers.

In 1988 the CEGB announced that Drax power station in Yorkshire, Western Europe's largest coal-fired power station, would be fitted with an FGD plant. A flow diagram is shown in Figure 3.6G. The £400 million FGD plant removes 90% of the sulphur dioxide discharged from the power station. The power station at Fidler's Ferry in Cheshire will also be fitted with an FGD plant. The calcium sulphate (gypsum) produced in the process will be sold to the plaster board industry and cement manufacturers. Should there be a surplus, it will be used for land-filling and reclamation of quarries, gravel pits and open-cast workings. The Drax FGD plant came into operation in stages between 1993 and 1995.

FGD plants can be fitted on to existing power stations. The UK is investing in FGD.

Flue gas desulphurisation is easier for the smelting industries, where the concentration of sulphur dioxide in the gases can be as high as 5–15%. Direct conversion into sulphuric acid is possible. This is a saleable commodity. If there is not a ready market, it can be converted into sulphur which is easier to store and to transport.

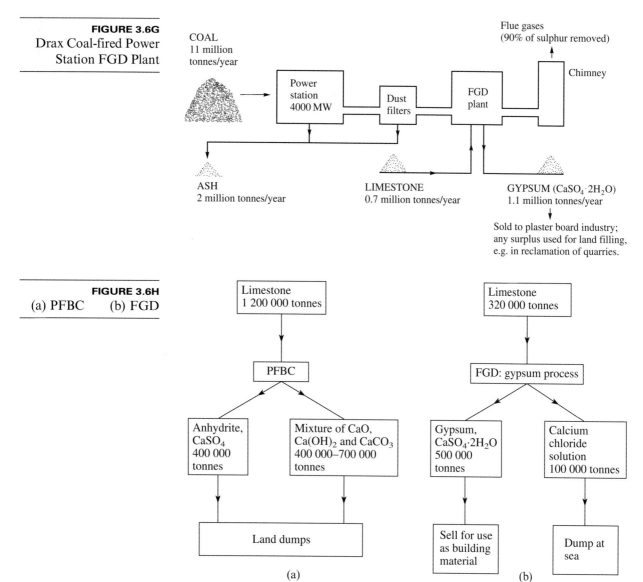

FIGURE 3.6G
Drax Coal-fired Power Station FGD Plant

FIGURE 3.6H
(a) PFBC (b) FGD

OTHER SOLUTIONS

Nuclear power stations and renewable energy sources are other options for reducing acid rain.

Other solutions to the problem of pollution from power stations are a switch to nuclear power stations and the development of technology for harnessing 'renewable energy sources, e.g. solar energy, wind, waves, hydropower and geothermal energy.

CHECKPOINT 3.6

1. (*a*) Why are oxides of sulphur air pollutants?

(*b*) How are they removed by natural processes?

2. Name (*a*) a natural source of nitrogen monoxide in the atmosphere, (*b*) a source from human activity.

3. Why is NO_2 more reactive than N_2O? What happens when N_2O reaches the stratosphere?

4. How do oxides of nitrogen contribute to acid rain?

5. The UK imports 750 000 tonnes of sulphur annually. The emission of sulphur from one power station is 1.3×10^6 tonnes of sulphur per year. Comment on these figures.

6. Refer to the comparison of PFBC and FGD in Figure 3.6H.

(*a*) Which process is better from the point of view of using limestone more carefully?

(*b*) Which process will obtain more energy from coal?

(*c*) How do the processes compare with respect to the disposal of waste products?

3.7 ORGANIC POLLUTANTS

Organic air pollutants have a serious effect on the quality of air. They have direct effects: for example, prolonged exposure to chloroethene causes cancer. They also form **secondary pollutants**, such as photochemical smog [see § 3.8]. In the case of hydrocarbons, the formation of secondary pollutants is the important effect. Organic pollutants are removed from the atmosphere by

Organic pollutants may have direct effects …
… or may form secondary pollutants …
… or may be removed from the atmosphere.

- precipitation in rainwater
- dry deposition
- photochemical reactions
- incorporation in particulate waste
- uptake by plants, especially trees

3.7.1 HYDROCARBONS

Hydrocarbons in the atmosphere include …

Hydrocarbons are widely used as fuels and may enter the atmosphere directly or as by-products of the partial combustion of other hydrocarbons. The hydrocarbons in uncontrolled vehicle exhausts are $\frac{1}{3}$ alkanes, $\frac{1}{3}$ alkenes and $\frac{1}{3}$ aromatic hydrocarbons. They are more reactive than hydrocarbons from other pollutant sources.

… methane, which comes from natural sources, contributes to the formation of ozone in the troposphere, and is oxidised to greenhouse gases …

Methane is produced in large quantities from the anaerobic decomposition of organic matter in water, sediments and soil. A source of 85 million tonnes of methane a year is the flatulent emissions from domestic animals which arise from bacterial digestion of food in the gut. Intensively cultivated rice fields produce as much as 100 million tonnes a year. Methane in the troposphere contributes to the photochemical production of carbon monoxide and ozone. Photochemical oxidation of methane is a major source of the greenhouse gases carbon dioxide and water vapour.

… and aromatic hydrocarbons from petrol and from industrial sources.

Aromatic hydrocarbons in the atmosphere include benzene, ethylbenzene and phenylethene. Some enter the atmosphere from petrol, some from industrial use as solvents and raw materials. Many plants emit unsaturated hydrocarbons called **terpenes**. They react very rapidly with $\cdot$ OH radicals and other oxidising agents including ozone.

3.7.2 ALDEHYDES AND KETONES

Aldehydes and ketones are emitted by industrial sources.

Aldehydes and ketones enter the atmosphere from industrial sources where they are used as solvents and as raw materials, and also from the internal combustion engine, incinerators and spray painting. Methanal and ethanal are produced by micro-organisms.

Carbonyl compounds are serious pollutants because the carbonyl group readily absorbs light in the near-UV region of the spectrum and is photolysed into a $\cdot$ CHO radical or a RCO$\cdot$ radical and an alkyl radical.

$$CH_3CHO + hv \longrightarrow CH_3 \cdot + \cdot CHO$$

$$RCOR' + hv \longrightarrow RCO \cdot + R' \cdot$$

They contribute to the formation of photochemical smog.

The radicals take part in the formation of **photochemical smog** [see § 3.8]. Unsaturated aldehydes are especially reactive in the atmosphere. Propenal, $H_2C{=}CHCHO$, also called acrolein, is a powerful **lachrymator** (tear producer) which is used as an industrial chemical and is a by-product of combustion.

3.7.3 HALOGEN COMPOUNDS

Halogen compounds include CFCs and halons, which attack the ozone layer.

Chlorofluoroalkanes or chlorofluorocarbons, known as **CFCs** or **freons**, e.g. CCl_2F_2, are used as aerosol propellants, refrigerants and foaming agents. Their effect on the ozone layer is discussed in § 3.11. **Halons**, fully halogenated alkanes containing bromine and fluorine, e.g. $CBrClF_2$, $CBrF_3$, $C_2Br_2F_4$, are used as fire-extinguishers. Their effect on the ozone layer is discussed in § 3.11.

3.8 PHOTOCHEMICAL SMOG

When atmospheric conditions result in a temperature inversion, pollutants remain near the ground instead of dispersing.

The movement of air across Earth's surface is a crucial factor in the dispersal of air pollutants. When air movement stops there can be a build-up of pollutants over a localised region. The temperature of air normally decreases with increasing altitude. Hot air rises, carrying pollutants with it, and is replaced by cooler air sinking to take its place. Sometimes atmospheric conditions can result in the reverse situation: increasing temperature with increasing altitude. This is known as a **temperature inversion**. It limits the circulation of the air, resulting in stagnation and the trapping of air pollutants near the ground [see Figure 3.8A].

FIGURE 3.8A
Pollutants Trapped in a
Temperature Inversion

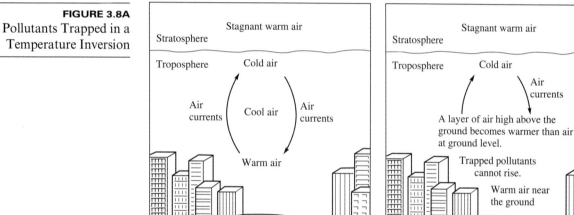

Hydrocarbons and nitrogen oxides react in UV light to form photochemical oxidants which create photochemical smog.

The pollution we call **photochemical smog** was first observed in Los Angeles, USA, in the 1940s. **Smog** is a yellowish haze which reduces visibility and causes eye irritation. The three factors needed to generate photochemical smog are ultraviolet light, hydrocarbons and nitrogen oxides. It forms more easily in stagnant air masses. **Photochemical oxidants** are formed and oxidise hydrocarbons to a mixture of compounds which form an aerosol haze. The oxidants include ozone, hydrogen peroxide, hydroxyl radicals, organic peroxides and hydroperoxides, ROOR' and ROOH, and peroxyacyl nitrates $RC(O)OONO_2$ (referred to as PAN).

Smog is a mixture of smoke, fog and sulphur dioxide.

The word **smog** is used to describe the combination of smoke and fog and sulphur dioxide that occurred frequently in London when coal with a high sulphur content was the principal fuel. The presence of sulphur dioxide makes it a **reducing smog**. In a photochemical smog sulphur dioxide is readily oxidised.

In photochemical smog, sulphur dioxide is readily oxidised.

The concentrations of the oxides of nitrogen, oxidants and hydrocarbons that take part in the formation of a photochemical smog vary throughout the day [see Figure 3.8B].

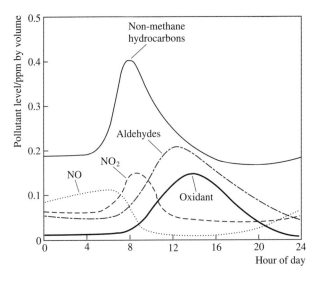

*Levels of pollutants vary
throughout the day.*

*The steps in the formation
of photochemical smog
are:*

The steps in the formation of a photochemical smog are:

1. NO_x and hydrocarbons are emitted into the atmosphere from early morning traffic and accumulate. After sunrise, NO_2 absorbs sunlight and photolyses.

*1. emission of NO_x and
hydrocarbons from
vehicles*

$$NO_2 \cdot + h\nu \longrightarrow NO \cdot + O \cdot$$

2. There is a build-up of oxidants: O_3, $O\cdot$, peroxides, excited $O\cdot^*$ atoms and hydroxyl radicals, $HO\cdot$. (In the following equations, RH = hydrocarbon and M is a third body which absorbs energy. All species are in the gas phase.)

*2. absorption of UV light
with the formation of
photochemical oxidants*

$$NO \cdot + O \cdot + M \longrightarrow NO_2 \cdot + M^*$$

$$O \cdot + O_2 + M \longrightarrow O_3 + M^* \text{ (a fast reaction)}$$

$$O \cdot + NO_2 \cdot \longrightarrow NO \cdot + O_2$$

$$O_3 + NO \cdot \longrightarrow NO_2 \cdot + O_2 \text{ (a fast reaction)}$$

This last reaction is fast; therefore $[O_3]$ remains low until $[NO\cdot]$ falls to a low value.

Hydroxyl radicals are formed by reactions which include the reaction of excited oxygen atoms with water and the photolysis of hydrogen peroxide:

$$O \cdot^* + H_2O \longrightarrow 2HO \cdot$$

$$H_2O_2 + h\nu \longrightarrow 2HO \cdot$$

The reactions of $HO\cdot$ include the attack on NO_x and on CO:

$$HO \cdot + NO_2 \cdot \longrightarrow HNO_3$$

$$HO \cdot + NO \cdot + M \longrightarrow HNO_2 + M^*$$

$$CO + HO \cdot + O_2 \longrightarrow CO_2 + HOO \cdot$$

The nitric acid and nitrous acid formed contribute to acid rain. The last reaction is a significant method for removing CO from the atmosphere and also a source of the reactive peroxyl radical $HOO\cdot$ which oxidises NO to NO_2 (in Step 4).

3. Oxidants react with hydrocarbons to form organic free radicals, e.g.

*3. reactions of oxidants
with hydrocarbons to
form free radicals*

$$O_3 + RH \longrightarrow R \cdot + RCO_2 \cdot + RCHO \text{ (or } R_2CO) + \text{Other products}$$

$$HO \cdot + RH \longrightarrow R \cdot + H_2O$$

$$R \cdot + O_2 \longrightarrow ROO \cdot \text{ (a strongly oxidising radical)}$$

A variety of compounds are formed, including aldehydes and ketones which condense to form aerosols which limit visibility. Oxidation of SO_2 to sulphuric acid occurs slowly in a clean atmosphere but much more rapidly (up to 50 times faster) in smoggy conditions.

4. Chain propagation occurs by a variety of reactions, e.g.

4. chain propagation

$$HOO\cdot + NO\cdot \longrightarrow HO\cdot + NO_2$$

$$NO\cdot + ROO\cdot \longrightarrow NO_2 + \text{other products}$$

These reactions explain why the oxidation of NO to NO_2 occurs so much more rapidly in an atmosphere in which NO_2 is undergoing photolysis.

5. Chain termination occurs when peroxide radicals react with NO_2 to form compounds such as **peroxyacyl nitrates**, **PAN**, e.g. peroxyethanoyl nitrate, $CH_3C(O)OONO_2$, and **peroxybenzoyl nitrate** $C_6H_5C(O)OONO_2$, **PBN**. These compounds are very strong eye irritants; they are very toxic to plants and they raise the overall level of oxidant.

5. chain termination with the formation of irritant compounds such as PAN and PBN.

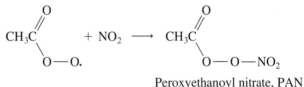

Peroxyethanoyl nitrate, PAN

The kinetics of the photo-oxidation of hydrocarbons in polluted air (Step 3) are very complex. Many steps are involved, and the picture given here is a very simplified version.

FIGURE 3.8C
The Formation of
Photochemical Smog

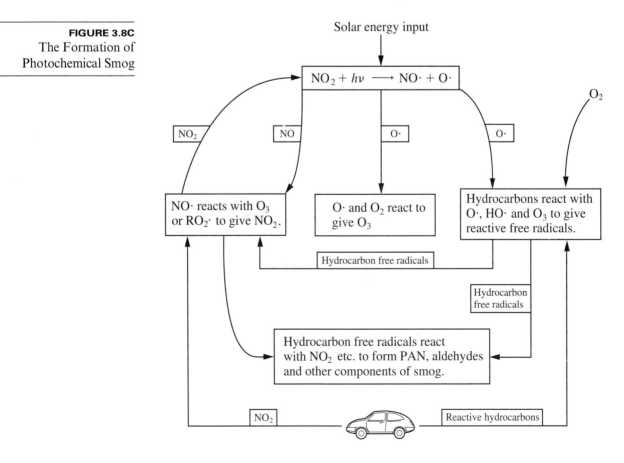

FIGURE 3.8D
Photochemical Smog

3.8.1 OXIDANT LEVEL

The total oxidant level can be found by oxidising iodide ion to iodine and measuring the iodine concentration by visible–UV spectrophotometry.

The **total oxidant level** can be determined by bubbling polluted air through aqueous potassium iodide to oxidise iodide ion to iodine, which then forms tri-iodide ion which can be determined in a visible–UV spectrophotometer from its absorption at 352 nm.

$$O_3(g) + 3I^-(aq) + 2H^+(aq) \longrightarrow O_2(g) + I_3^-(aq) + H_2O\ (l)$$

Ozone levels can be measured by a chemiluminescent method, in which ozone reacts with dyes such as rhodamine B and the intensity of the light emitted is measured.

PAN is measured by IR spectrometry.

PAN can be measured by infrared spectroscopy. PAN, aldehydes and ketones can be detected and measured by gas chromatography, and by gas chromatography linked to a mass spectrometer as a detector.

3.8.2 THE EFFECTS OF PHOTOCHEMICAL SMOG

The effects of photochemical smog can be divided into:

1. effects on human health

2. damage to materials

3. effects on the atmosphere

4. toxicity to plants

Photochemical smog irritates the respiratory system ...

1. Ozone at 0.15 ppm causes coughing, wheezing and irritation to the respiratory systems of healthy individuals. Some people are more susceptible than others. In tests on laboratory animals ozone has been shown to lower resistance to disease. PAN and aldehydes are eye irritants.

2. Ozone causes rubber to deteriorate through fission of the double bond.

... attacks rubber, dyes and fabrics ...

The damage can be reduced by the use of ozone-resistant rubbers such as poly(chloropropene). The attack of ozone on double bonds also reduces the strength of fabrics and causes bleaching of dyes.

3. Aerosol particles that reduce visibility are formed by polymerisation of smaller particles produced in photochemical smog-forming reactions. These reactions involve oxidation of hydrocarbons, therefore oxygen-containing organic compounds make up the bulk of particulate matter produced by smog.

... reduces visibility ...
... and damages crops.

4. There is a shortage of food worldwide, and the effects of photochemical smog on plants is therefore important. The effects are due to the oxidants PAN, ozone and NO_x. Injury may occur after exposure to 0.03 ppm ozone for 8 hours. Shorter times are required if sulphur dioxide is also present. Ozone reduces the rate of photosynthesis, producing yellow spots on the leaves and reducing plant growth. PANs are very toxic, attacking leaves and causing bronzing and glazing of their surfaces. Damage can occur after exposure to 0.01 ppm PAN for 5 hours. NO_x have relatively low toxicity. In California, crop damage from ozone and PAN are estimated to cost millions of dollars a year.

CHECKPOINT 3.8

1. Which of the following species reaches its peak last in a smoggy atmosphere?

A NO,   **B** oxidants,   **C** hydrocarbons,   **D** NO_2

2. Name a species that oxidises NO to NO_2 in a smoggy atmosphere.

3. How can oxidants be detected in the atmosphere?

4. Why is ozone damaging to rubber?

5. Hydrocarbons are not toxic, so why are they considered to be serious pollutants?

6. Name a substance which is present in a reducing smog but not in a photochemical smog.

7. Which of the following is least likely to be formed when a photon of light is absorbed by a molecule of NO_2?

A O atom,   **B** NO,   **C** NO_2*,   **D** N atom

8. (*a*) What kind of weather favours the formation of photochemical smog?

(*b*) What kind of geographical locations are prone to photochemical smog?

(*c*) Explain why the conditions you mention in (*a*) and (*b*) favour the formation of photochemical smog.

3.9 VEHICLE ENGINES

The formation of photochemical smog [§ 3.8] arises from nitrogen monoxide and hydrocarbons. The major source of both of these is the exhaust gases of vehicles with internal combustion engines. The world has 200 million motor vehicles which cover 2 million million miles a year and burn 70 billion litres of petrol. From their exhausts come 40 million tonnes of carbon monoxide, 4 million tonnes of oxides of nitrogen and 4 million tonnes of hydrocarbons. Until the introduction of unleaded petrol, 0.2 million tonnes of lead were part of this discharge.

Vehicle engines are the source of the pollutants: carbon monoxide, oxides of nitrogen, hydrocarbons and lead compounds.

The pollutants are discharged at street level where people cannot avoid inhaling them. Carbon monoxide is a poisonous gas [see § 3.3]. Hydrocarbons by themselves cause little damage but in the presence of sunlight they form photochemical smog [see § 3.8]. Nitrogen oxides play an important part in the formation of photochemical smog [see § 3.8], the depletion of the ozone layer [see § 3.11] and the formation of acid rain [see § 3.6].

3.9.1 CONTROLLING THE POLLUTANTS

In the petrol engine, a mixture of fuel and air is ignited in each cylinder by a spark. The burning is sudden and intense. The temperature soars to 2800 °C, and some nitrogen and oxygen in the cylinder combine to form nitrogen monoxide. As the piston is pushed out of the cylinder, the combustion gases expand and cool in under one hundredth of a second. The cycle of heating and cooling occurs so rapidly that some of the hydrocarbons in the fuel remain uncombusted and some are incompletely oxidised to carbon monoxide. The simultaneous production of oxidised contaminants, e.g. oxides of nitrogen, NO$_x$, and incompletely oxidised contaminants, e.g. hydrocarbons and carbon monoxide, makes the problem of cleaning up the exhaust a difficult one to solve.

Adjustments can be made to the engine: to the fuel supply, air/fuel ratio and spark timing. These adjustments are called **tuning**. The problem in tuning the engine to reduce emissions is that the composition of the exhaust changes according to driving conditions [see Table 3.9A].

Exhaust component/%	*Driving mode*			
	Idling	*Cruising*	*Accelerating*	*Decelerating*
Carbon monoxide	5.2	0.8	5.2	4.2
Hydrocarbons	0.075	0.030	0.040	0.40
Nitrogen oxides	0.0030	0.15	0.30	0.0060

In the past, engines were tuned to achieve peak performance. Now that the public are concerned about the quality of the environment, manufacturers have to balance performance against pollution.

The equation for the complete combustion of octane is:

$$2C_8H_{18} + 25O_2 \longrightarrow 16CO_2 + 18H_2O$$

From the stoichiometry of the equation it follows that (mass of air/mass of octane) = 14.7. This is called the **air/fuel ratio**. According to the composition of the petrol, an air/fuel ratio of 14 or 15 will give complete combustion. Figure 3.9A shows how the production of each major pollutant depends on the air/fuel ratio.

A rich mixture (with less than the stoichiometric proportion of air) gives an exhaust gas which is high in CO, high in hydrocarbons and low in NO$_x$. A lean mixture (with an excess of air) gives an exhaust gas with less CO and less hydrocarbon and reaches a peak in NO$_x$ at an air/fuel ratio of about 15.5. A rich mixture is better for smooth engine performance. With a very lean mixture the engine misfires and the level of hydrocarbon rises. Changing the air/fuel ratio does not solve the problem: it merely trades one set of pollutants for another.

The combustion of petrol vapour inside the cylinders of a car engine produces a large volume of hot gases. The gases force the piston down the cylinder and transmit power to the wheels. For smooth running it is essential that ignition of petrol vapour in air takes place when the piston is at the right point in the cylinder. If the petrol vapour ignites before the ignition spark, the pressure increases suddenly and the piston makes a metallic sound called knocking. The fuel 2,2,4-trimethylpentane, $(CH_3)_2CHCH_2C(CH_3)_3$, formerly called *iso*-octane, has a good resistance to knocking and is assigned an **octane number** of 100. Heptane, $CH_3(CH_2)_5CH_3$, has bad knocking properties and has an octane number of 0. Other fuels are assigned octane numbers on the same scale.

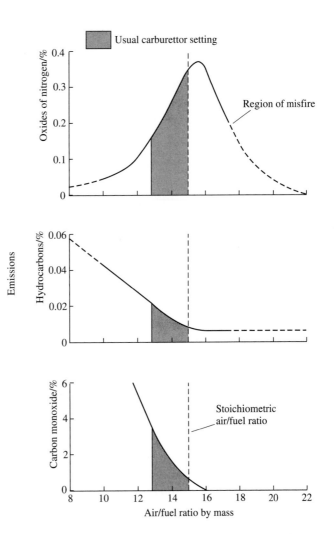

FIGURE 3.9A
Air/fuel Ratio and
Exhaust Composition

The octane number of a fuel can be increased by the addition of 0.5–1.0 g of lead per litre. This is added as **tetraethyllead**, **TEL**, $Pb(C_2H_5)_4$. TEL has been an important additive in petrol for 50 years. During combustion TEL reacts with other additives to form volatile lead compounds which escape with the exhaust gases. Lead compounds are toxic [see § 3.2]. Concern over the release of lead compounds into the air has led to the banning of TEL in a number of countries. Since 1990 all new vehicles in the UK have been designed to run on **unleaded petrol**.

The petrol used in cars made since 1990 does not include tetraethyllead.

One solution to the problem of pollution by vehicles is to find alternative fuels with harmless combustion products [see § 3.10]. Other solutions are the catalytic converter [§ 3.9.2] and improved engine design, of which the lean-burn engine is one example [§ 3.9.3].

3.9.2 CATALYTIC CONVERTER

Both carbon monoxide and hydrocarbons can be rendered harmless by oxidation. This can be achieved by passing the exhaust gases mixed with air through a **catalytic converter**. This is a pipe containing a solid catalyst, platinum and rhodium, which catalyses the oxidation of carbon monoxide to carbon dioxide and hydrocarbons to carbon dioxide and water [see Figure 3.9C]. It cannot remove NO_x by oxidation. The catalyst is inactivated by lead and therefore requires unleaded petrol. Catalytic converters have been fitted to all new UK vehicles since 1990 and can be retrofitted to older vehicles.

Unleaded petrol must be used in vehicles fitted with catalytic converters.

FIGURE 3.9C
A Catalytic Converter

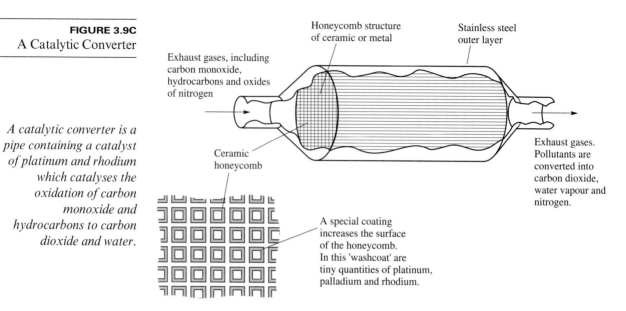

A catalytic converter is a pipe containing a catalyst of platinum and rhodium which catalyses the oxidation of carbon monoxide and hydrocarbons to carbon dioxide and water.

An engine can be manufactured with a **three-way catalytic system**, which can both oxidise hydrocarbons and carbon monoxide and also reduce NO_x. The exhaust gas is passed over a reduction catalyst of e.g. platinum, palladium, cobalt or nickel. Carbon monoxide acts as the reducing agent:

$$2NO(g) + 2CO(g) \longrightarrow N_2(g) + 2CO_2(g)$$

The harmless products nitrogen and carbon dioxide are formed. Air is taken in downstream from the reduction catalyst, and the exhaust gas and air pass over an oxidation catalyst of e.g. palladium, platinum or rhodium to oxidise hydrocarbons and carbon monoxide.

$$2C_8H_{18}(g) + 25O_2(g) \longrightarrow 16CO_2(g) + 18H_2O(g)$$

$$2CO(g) + O_2(g) \longrightarrow 2CO_2(g)$$

A three-way catalytic system reduces NO_x at the same time.

It is not possible to fit a three-way system to an existing car.

3.9.3 LEAN-BURN ENGINES

Another solution to the problem of cleaning up vehicle exhaust gases is the **lean-burn engine**. The stoichiometric air/fuel ratio is 14 : 7 by mass. Lean-burn engines use an air/fuel ratio of about 18 : 1. The tendency is for the engine to misfire on a lean mixture and it must be modified. The mixture of fuel and air is electronically controlled. A rich fuel–air mixture is injected into a pre-chamber, in which it ignites easily. The main region of the cylinder contains the lean mixture. It is ignited by the flame in the pre-chamber. A feature of lean-burn engines is that they produce less NO_x than traditional engines [see Figure 3.9A]. The lean-burn engine gives less pollution and better fuel economy.

Lean-burn engines use a high air/fuel ratio. They produce less NO_x and have better fuel economy.

CHECKPOINT 3.9

1. (*a*) Explain why the concentration of carbon monoxide in exhaust gases is:

(i) higher during acceleration than during cruising

(ii) higher during idling than in cruising.

(*b*) Explain why the concentration of hydrocarbons in exhaust gases is higher during deceleration than during acceleration.

(*c*) Explain why the concentration of nitrogen monoxide is higher during acceleration than during idling.

2. Give three advantages of running cars on unleaded petrol.

3. A car which was new in 1980 drove 100 000 miles before it was scrapped. It ran on three-star petrol containing 3.5 g of TEL per gallon, and it did 35 miles to the gallon. How much TEL did this car use in its lifetime? What happened to the lead?

4. It has been suggested that battery-operated vehicles are the solution to vehicle pollution problems. What effect would the increased use of battery-operated cars have on the demand for electricity? What would be the consequences for the environment of this demand?

3.10 ALTERNATIVE FUELS

Fuels used in the internal combustion engine must vaporise at the engine temperature, must dissolve in petrol, and must form harmless combustion products.

Countries which have to import oil are interested in finding alternatives. The major use of petroleum oil fractions is in vehicle engines. The use of alternative fuels would extend the life of Earth's finite resources of oil and also allow crude oil to be used in the petrochemicals industry. Petrol engines are designed to operate over a temperature range at which petrol will vaporise. Any fuel added to petrol must vaporise at the engine temperature, must dissolve in petrol and must form harmless combustion products. The commonest substances which are blended with petrol are listed in Table 3.10A with straight-run gasoline, which has not been improved by cracking, reforming, etc.

Substance	Formula	Octane no.	Boiling temperature/ $^\circ$C	Price/ p l^{-1}
Ethanol	C_2H_5OH	111	79	12.0
Methanol	CH_3OH	114	65	4.0
MTBE	$CH_3OC(CH_3)_3$	118	55	10.0
Gasoline (straight run)		70		5.0

TABLE 3.10A

The substances are referred to as **oxygenates** in the oil industry. They increase the octane number of petrol and cause less pollution, especially in the emission of carbon monoxide.

3.10.1 ETHANOL

Ethanol burns well in petrol engines as mixtures containing up to 10% ethanol.

During the Second World War, Germany was short of petroleum oil and many vehicles were adapted to run on ethanol which had been made from potatoes. Ethanol dissolves in petrol and it boils at the same temperature as heptane, 79 °C. Ethanol burns well in petrol engines, producing about 70% as much energy per litre as petrol. A petrol engine will take 10% ethanol in the fuel without any adjustments to the carburettor (which controls the ratio of air to fuel in the cylinders). Ethanol burns to form carbon dioxide and water and a little ethanal; there is little atmospheric pollution. A mixture of ethanol and petrol is described as **gasohol**.

Brazil makes ethanol from sugar cane, which grows well in sunny climates. Brazil clears forests to obtain more agricultural land.

Ethanol can be produced by the fermentation of sugars and starches. It is an attractive proposition for a country with no oil but plenty of arable land to grow crops for fermentation. Brazil has plenty of land on which to grow sugar cane and has the sunshine needed to ripen the crop. Most of the petrol sold in Brazil contains 10% ethanol, and this reduces the cost of oil imports. Brazil hopes to increase the content of ethanol in petrol in the future. To obtain more agricultural land, Brazil is clearing large areas of rainforest. You will be aware of the environmental cost in the loss of habitats for forest species, in the loss of trees which absorb carbon dioxide and maintain a steady level of carbon dioxide in the carbon cycle, and in the emission of carbon dioxide when the trees are burnt, thus contributing to the greenhouse effect.

FIGURE 3.10A
Petrol Pumps in Brazil
Selling 'Alcool' Gasohol

3.10.2 METHANOL

Methanol burns well in vehicle engines. It is made from methane and steam.

Methanol burns cleanly in a vehicle engine to form the harmless combustion products carbon dioxide and water, with little carbon monoxide emission. There is no emission of aromatic hydrocarbons such as the carcinogen, benzene. The octane number of methanol is high (114), and the carburettor requires only a small adjustment. Racing drivers use methanol as fuel because, being less volatile than petrol, it is less likely to explode in a collision.

Methanol is inexpensive because it can be made from a number of plentiful raw materials, including methane and coal. In the steam reforming process, methane and steam are used to make synthesis gas ($CO + H_2 + CO_2$), which is converted into methanol.

The co-solvent, e.g. TBA, which must be added to help methanol to mix with hydrocarbons makes the fuel corrosive and electrically conducting.

Methanol is used as a mixture with petrol in some countries, including Germany and parts of the USA. It does not mix well with hydrocarbons unless a co-solvent is added, such as $(CH_3)_3COH$, 2-methylpropan-2-ol (called TBA after its old name of tertiary butyl alcohol). The mixture is hygroscopic (absorbing water from the air). When the water content reaches 0.5% the fuel separates into a layer of petrol on top of a layer of methanol and water. The water content makes the fuel corrosive, and fuel tanks must be made out of stainless steel. It conducts electricity and electrical components of engines need to be well insulated. Methanol produces only 60% of the energy of an equal volume of petrol, so fuel tanks have to be larger, increasing the weight of the vehicle, or filled much more frequently.

Methanol vapour and methanal, a product of incomplete oxidation, are toxic.

Methanol is a toxic substance. It is very dangerous to drink methanol, and prolonged exposure to the vapour also has a damaging effect on the brain. There is concern that petrol pump attendants and mechanics would be exposed to the vapour for many hours and would suffer damage. The exhaust fumes of cars driven on methanol contain methanal, which is lachrymatory and carcinogenic.

3.10.3 MTBE

The fuel MTBE is made from methanol.

Another contribution of methanol to alternative fuels is as a source of 2-methoxy-2-methylpropane, $(CH_3)_3COCH_3$, known as MTBE after its old name of methyl tertiary butyl ether. It is manufactured from methanol and 2-methylpropene:

$$CH_3OH(g) + (CH_3)_2C = CH_2(g) \longrightarrow (CH_3)_3COCH_3(g)$$

MTBE is more expensive than petrol and is added up to 7% by volume to increase the oxidation number and reduce the emission of carbon monoxide.

MTBE is another oxygenate. With octane number 118, MTBE is blended with petrol up to 15% by volume to increase the octane number and reduce the emission of carbon monoxide. More than 15% by volume of MTBE (b.p. 55 °C) would make the mixture too volatile. In practice, the addition is limited to 0.6% in unleaded regular, 2.2% in unleaded premium and 7% in unleaded super plus. MTBE costs more than gasoline. The cost of using a high percentage of MTBE has to be weighed against the cost of increasing the octane number of gasoline by isomerising, cracking and reforming the alkanes in it. Straight-run gasoline has an octane number of about 70. This can be increased to about 80 by isomerisation, to about 90 by catalytic cracking and to about 95 by reforming [see *ALC*, §§ 26.3.1–4].

3.10.4 HYDROGEN

Hydrogen is a clean fuel, readily available, but impure hydrogen burns explosively in air and it must be used with great care.

Hydrogen is a 'clean' fuel, burning to form water vapour only. The internal combustion engine can be converted to run on hydrogen by a modification to the carburettor. The fuel storage tank is the main problem: impure hydrogen burns explosively in air and a leak would be very dangerous. Hydrogen is more suited for aircraft than for automobiles. Better safety precautions can be taken over filling the tank. Hydrogen is an ideal fuel for gas turbine engines, and the smaller mass of fuel that must be carried, compared with kerosene, is an advantage.

CHECKPOINT 3.10

1. (*a*) What pollution problem in the internal combustion engine is aggravated by a lean mixture?

(*b*) Outline the principles behind a lean-burn internal combustion engine.

(*c*) Why are two catalytic reactors necessary to control all the major exhaust pollutants from a motor vehicle?

2. Review the advantages of using ethanol as a fuel for cars. In which type of country is there special interest in this fuel?

3. Review the advantages and disadvantages of methanol as a fuel for cars.

4. (*a*) What is meant by the statement 'Hydrogen is a clean fuel'?

(*b*) What is the drawback of hydrogen as a fuel for cars?

3.11 OZONE IN THE STRATOSPHERE

The ozone layer keeps the Earth at a temperature that supports life.

The ozone layer in the stratosphere was described in §§ 2.3, 2.4, 2.7 and 2.10. The vital importance of the ozone layer is that it protects the Earth from harmful ultraviolet radiation and makes it a place where animals and plants can live. Pollutants from human activities are affecting the ozone layer.

3.11.1 CHLORINE COMPOUNDS IN THE STRATOSPHERE

During the 1970s people began to consider the effect of chlorine compounds on ozone in the stratosphere. Chloromethane is the source of about 25% of the chlorine in the stratosphere. Most of it is biological in origin, e.g. the rotting of wood, and it totals about 5.5 million tonnes a year. Some hydrogen chloride is formed by hydrolysis of chloromethane and reacts to give Cl· atoms which react with ozone. Chlorine is even more effective than oxides of nitrogen in removing ozone.

Chlorine compounds in the stratosphere are photolysed to give chlorine atoms ...

Initiation step is:

$$\cdot OH + HCl \longrightarrow H_2O + Cl\cdot$$

Propagation steps are:

(a) $Cl\cdot + O_3 \longrightarrow ClO\cdot + O_2$

This reaction is 2000 times faster than the reaction between $NO\cdot$ and O_3.

(b) $ClO\cdot + O\cdot \longrightarrow Cl\cdot + O_2$

(c) $O_3 + ClO\cdot \longrightarrow O_2 + ClO_2$

Note that reactions (a) and (b) constitute a chain reaction. The chain length is given by:

... which react with ozone in a chain reaction with a chain length of 5000.

$$\text{Chain length} = \frac{\text{Propagation rate}}{\text{Termination rate}}$$

In this case, chain length = 5000, that is, one chlorine atom can destroy about 5000 molecules of ozone.

Termination steps are:

$$Cl\cdot + CH_4 \longrightarrow HCl + \cdot CH_3$$

$$Cl\cdot + H_2 \longrightarrow HCl + H\cdot$$

$$ClO\cdot + NO_2\cdot + M \longrightarrow ClNO_3 + M^*$$

$ClNO_3$ is a temporary reservoir for chlorine. It can photolyse to give $Cl\cdot$ atoms.

3.11.2   CHLOROFLUOROCARBONS, CFCs

Chlorofluorocarbons, CFCs, are chemically stable, non-toxic compounds which have found wide use as refrigerants, as aerosols, solvents, etc.

In the 1970s the importance of chlorofluoroalkanes, known as **chlorofluorocarbons, CFCs**, was realised. CFCs are chemically stable, of low toxicity, are volatile liquids of low cost which are ideal for use as refrigerants, as blowing agents in plastic foams, and as aerosols, as solvents and cleaning agents in the electronics industry. The chief CFCs in use are $CFCl_3$, CF_2Cl_2 and $CF_2ClCFCl_2$. Although CFCs are denser than air, they diffuse slowly into the stratosphere. Being chemically unreactive, CFCs pass through the troposphere to the stratosphere where they are photolysed at altitudes of 20–49 km to give chlorine atoms which can attack the ozone layer as described above.

$$CCl_2F_2 + hv \longrightarrow \cdot CClF_2 + Cl\cdot$$

$$CCl_3F + hv \longrightarrow \cdot CCl_2F + Cl\cdot$$

They diffuse slowly into the stratosphere where they are photolysed to give chlorine atoms which attack the ozone layer.

Even if CFC production were stopped immediately, CFCs already in the troposphere would affect the ozone layer for several decades. There are no known sinks in the troposphere and the residence time is over 60 years. There is no natural source of CFCs. In 1988 one million tonnes were produced.

3.11.3   EVIDENCE FOR OZONE DEPLETION

During the 1970s, a team of British scientists led by Dr Joe Farman were working at the British Antarctic Survey Station. They were using spectrometry to measure the composition of the atmosphere. They were amazed to find a big decrease, about 20%, in the ozone layer over the Antarctic. This was all the more amazing because the US National Aeronautics and Space Administration, NASA, had been monitoring the region by means of an orbiting satellite for many years. When the British team

Since 1970, evidence has accumulated of a thinning of the ozone layer over the Antarctic: the 'ozone hole' ...

announced their discovery, the Americans re-examined their records. The NASA computers had been programmed to record small changes in the ozone layer and to dismiss any changes greater than 30% as erroneous readings. When the NASA records were re-examined they were found to support the British observations. There is a large decrease in ozone over Antarctica during the spring (August–October), which is described as the **ozone hole**. Later in the year the level recovers. Why does the ozone hole appear over the Antarctic? The temperature of the stratosphere there is low enough for clouds to form, and reaction at the surface of water droplets increases the rate of decomposition of 'chlorine reservoirs' of $ClNO_3$ to give $ClO\cdot$. Measurements show about 100 times the concentration of $ClO\cdot$ in the Antarctic compared with temperate latitudes. Why does the ozone hole appear in the spring? The clouds warm up and release $ClO\cdot$ radicals. Why is the phenomenon largely confined to the Antarctic? The air mass over the Antarctic, called the polar vortex, rotates in isolation from the rest of the global air mass.

... which is greatest in the Antarctic spring and recovers later in the year.

In March 1989, a team of scientists working in the Canadian Arctic detected a thinning of the ozone layer over the Arctic. In 1990 sampling by a high-altitude US plane showed that the levels of ozone-depleting gases are 50 times higher than expected over the Arctic. An ozone 'hole' over the Arctic would be even more serious than the 'hole' over the Antarctic because some of the most populated parts of the globe, including North America, northern Europe, Russia and associated countries will have less protection from UV radiation. The 'hole' over the Arctic is increasing more slowly than that over the Antarctic because the temperature in the Arctic is higher and the ozone-destroying chemicals are less effective.

Since 1989, a thinning of the ozone layer over the Arctic has been detected. This is closer to populated parts of the world than is the Antarctic 'hole'.

FIGURE 3.11A
The Ozone 'Hole' over the Antarctic

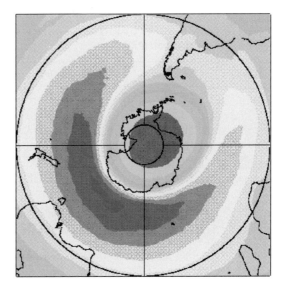

3.11.4 CONSEQUENCES OF OZONE DEPLETION

A thinning of the ozone layer would increase the intensity of UV light reaching Earth's surface. This would damage crops ...
... and phytoplankton, thus leading to a decrease in fish stocks.

Increases in the intensity of ultraviolet radiation reaching the troposphere would have serious consequences. One major effect would be on plants, including crops used for food. Aquatic organisms are very sensitive to small increases in UV radiation. The destruction of phytoplankton, the microscopic plants that are the basis of the ocean's food chain, would disrupt the food chain, affecting yields of fish which are an important part of the human diet.

The major effect of ultraviolet radiation on human beings occurs in the skin. Nucleic acids are the main absorbers and changes in DNA can result in uncontrolled cell division leading to skin cancer. This is generally a non-melanoma, a non-fatal cancer, but there is evidence of a connection between UV radiation and melanoma skin

An increase in UV radiation would increase skin cancers ...

... and cause cataracts and other eye damage ...

... and lower resistance to disease.

cancer, which is a more frequently fatal form of the disease. The US Environmental Protection Agency has predicted that a 10% reduction in the ozone layer would result in a 20% increase in both forms of skin cancer. This would mean 16 000 extra cases of non-melanoma skin cancers per year in the USA and 8000 in the UK. Excessive UV radiation can also cause cataracts and other eye damage, and it lowers resistance to disease. One estimate is that a 1% decrease in the ozone layer will cause 20 000 premature deaths over the lifetime of the existing US population.

3.11.5 ALTERNATIVES TO CFCs

Hydrohalocarbons contain at least one C—H bond per molecule ...

To preserve the ozone layer, we must discontinue our use of CFCs. Chemists have proposed alternatives which are already in production. Halogenoalkanes that contain at least one hydrogen atom per molecule are known as **hydrohalocarbons**. The C—H bond is susceptible to attack by HO· radicals in the troposphere, and the compounds do not reach the stratosphere. Hydrohalocarbons include:

... e.g. hydrochlorofluorocarbons, HCFCs, such as $CHCl_2CF_3$ *...*

● **hydrochlorofluorocarbons**, **HCFCs**, hydrogen-containing chlorofluoroalkanes, e.g. $CHCl_2CF_3$ used in plastic foam-blowing, $CHClF_2$ used in air conditioners and plastic foam food containers

... e.g. hydrofluorocarbons, HFCs, such as CH_2FCF_3.

● **hydrofluorocarbons**, **HFCs**, hydrogen-containing fluoroalkanes, e.g. CH_2FCF_3 used in air conditioners and refrigeration equipment.

They are alternatives to CFCs ...

The lifetimes of HCFCs and HFCs depend on the molar mass, the number of hydrogen atoms in the molecule and the number of chlorine and fluorine atoms adjacent to the hydrogen atoms. The ability to destroy ozone is expressed by the **ozone depletion potential**. Some values, relative to $CCl_3F = 1$ are given in Table 3.11A.

Compound	Ozone depletion potential
CCl_3F	1.0
$CHClF_2$	0.030
$CHCl_2CF_3$	0.013
CH_3CCl_2F	0.10
$CHClFCF_3$	0.035
CH_3CClF_2	0.038

TABLE 3.1A

HCFCs destroy one third of the amount of ozone destroyed by CFCs. HFCs cause no damage to the ozone layer, although they are greenhouse gases.

... as are perfluorocarbons, which are completely fluorinated alkanes, e.g. CF_4.

Another alternative to CFCs is **perfluorocarbons** which are completely fluorinated alkanes, e.g. CF_4 and C_2F_6. They do not react with hydroxyl radicals or ozone; they are not involved in ozone depletion or photochemical smog formation. Their lifetime has been estimated as tens of thousands of years, and they may add to greenhouse warming. Their very long lifetime makes this a serious threat.

3.11.6 POLITICS OF THE OZONE LAYER

Many nations have signed a treaty, agreeing to reduce their use of CFCs and to phase them out completely in the near future ...

In 1985, 49 countries signed the Vienna Convention for the Protection of the Ozone Layer. In 1987 the Montreal Protocol on Substances that Deplete the Ozone Layer amended the original treaty by agreeing a faster reduction in the use of CFCs. The consumption of CFCs was cut by 20% by mid-1993 and by 50% by mid-1996. The London meeting in 1990 of 124 nations who had signed the treaty agreed a total phase-out of CFCs by 1999 and a prudent use of HCFCs. At the 1992 Convention in Copenhagen, the industrial nations pledged to phase out HCFC production by 2020.

... and to reduce their use of HCFCs and phase them out eventually.

By the time that the 1995 Convention was held in Vienna, 150 governments had become party to the ozone treaty. In 1985 the world production of CFCs was 1 million tonnes a year. By 1995 the production was still 360 000 tonnes. The reason is that the London 1990 agreement allows developing countries to go on producing and importing CFCs until 2010. The poor nations, led by India and China, argue that the rich nations created the CFC problem, yet it is the rich nations who are telling the poor nations that they must use safer and more expensive alternatives to help to repair the damage which the rich nations caused. China used 80% more CFCs in 1995 than it did in 1993. Many newly industrialised countries have expanded their use of HCFCs

Many developing nations have not signed because they think it unfair that they should be asked to solve a problem that the rich nations have created.

The solution to the reluctance of poorer nations to switch to dearer alternatives seems to be for the richer nations to subsidise them.

as refrigerants. According to a study by Greenpeace, the refrigeration market in industrialised Asia is expected to grow as rapidly as it did in postwar Japan. With this growth, the potential for ozone destruction in 2030 would be twice that of 1990. Bromomethane is another ozone-depleting substance. It is used to fumigate soil and to fumigate vegetables and cut flowers for export. Western countries have phased it out and found substitutes. Developing countries say that some alternatives are too expensive for poor countries. China is about to expand its use of bromomethane.

The problem needs to be solved.

At the 1995 meeting the developing countries agreed to phase out the use of HCFCs by 2010 and to freeze the use of bromomethane by 2002 . The condition is that they receive money from the rich nations to pay for a switch to alternatives. The London meeting in 1990 set up a Multilateral Fund to help poor countries to finance a switch to safe alternatives. Unfortunately the fund is receiving only 85% of the money which the rich countries promised. Only if all nations comply with the terms of the 1992 Montreal Protocol will chlorine levels in the stratosphere decline to a safe level of 2 ppb some time in the middle of the twenty-first century.

3.11.7 DISPOSAL OF CFCS

The disposal of CFCs is by burning at high temperature, which is costly ...
.... or by a new method – passing through sodium ethanoate at 280 °C.

Organisations in industrialised countries that have stockpiles of CFCs are either selling them to countries which are still allowed to use them or releasing them into the air. A commercial method of breaking down CFCs is to burn them at high temperature, usually in cement kilns. This is so expensive that the Vienna congress in 1995 decided that it was impractical to demand that industries burn CFCs.

A recent discovery is that, when passed through sodium ethanedioate at 280 °C, CFCs break down to form sodium chloride, sodium fluoride, carbon and carbon dioxide. The technique can be adapted to produce unsaturated fluoro compounds which can be used in the synthesis of new compounds. The destruction of CFCs has suddenly become practicable.

CHECKPOINT 3.11

1. (*a*) What evidence is there that the ozone layer is thinner than it was?

(*b*) In what parts of the world is the effect most pronounced?

(*c*) What makes these regions particularly vulnerable to thinning of the ozone layer?

(*d*) Why is there so much concern over the ozone 'hole'?

2. (*a*) Write the equations for two reactions in which free radicals containing chlorine react with ozone.

(*b*) Why is one chlorine atom able to destroy about 5000 molecules of ozone?

(*c*) How are free radicals formed from chlorofluorocarbons, CFCs?

(*d*) Why do CFCs pass through the troposphere without reacting?

3. (*a*) Mention some of the purposes for which CFCs are used, and say why they are suited to these uses.

(*b*) What are the alternatives to CFCs?

4. (*a*) If all nations stopped using CFCs tomorrow, would the problem of CFCs be solved?

(*b*) Which nations are reluctant to stop using CFCs?

(*c*) Suggest how their reluctance could be overcome.

3.12 OZONE IN THE TROPOSPHERE

In the troposphere ozone is a pollutant ... ... which damages vegetation and impairs human health.

In the stratosphere ozone protects the Earth from the effects of receiving too much UV radiation. At ground level ozone is a toxic gas which can damage vegetation and impair human health. Ozone is a highly toxic, irritant gas with a pungent odour and a bluish colour at high concentration. It decomposes to form oxygen, with a half-life of 3 days at 20 °C. It is a powerful oxidising agent which is capable of reacting with many substances in the atmosphere.

The natural ozone level in the troposphere is 20–50 ppb. It has diffused from the stratosphere where it is formed from dioxygen through the absorption of radiation of wavelength < 242 nm [see §§ 2.7, 3.11]. It cannot be formed in the same way at ground level because light of wavelength < 290 nm is absorbed in the stratosphere by the ozone layer.

One substance which can be photolysed in the range of visible–UV radiation that penetrates to the lower atmosphere is nitrogen dioxide, NO_2. It is a secondary pollutant, derived from nitrogen monoxide, NO [see § 3.5]. Nitrogen dioxide photolyses to $O\cdot$, which reacts with O_2 to form O_3, and $NO\cdot$, which reacts with O_3 to reform NO_2.

Nitrogen dioxide photolyses to form free radicals which react with oxygen to form ozone.

(a) $NO_2\cdot + h\nu \longrightarrow NO\cdot + O\cdot$

(b) $O\cdot + O_2 + M \longrightarrow O_3 + M^*$

(c) $NO\cdot + O_3 \longrightarrow NO_2\cdot + O_2$

Reaction (b) produces ozone, while reaction (c) destroys ozone. A steady state is therefore reached.

Photochemical reactions of ozone result in the formation of hydroxyl radicals.

$$O_3 + h\nu \longrightarrow O\cdot + O_2$$

$$O\cdot + H_2O \longrightarrow 2\cdot OH$$

$\cdot$OH radicals can abstract H from hydrocarbons to form alkyl radicals.

$$\cdot OH + RCH_3 \longrightarrow H_2O + \cdot RCH_2$$

FIGURE 3.12A

Ozone Levels Throughout the Day

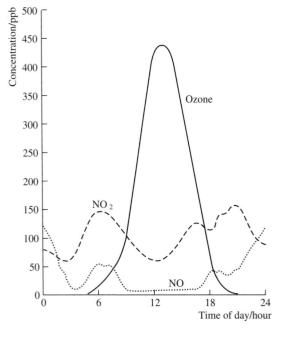

Ozone oxidises hydrocarbons in the formation of photochemical smog.

Alkyl radicals can add molecular oxygen to form peroxy radicals, $RCH_2O_2 \cdot$ which react with NO to form NO_2.

$$RCH_2O_2 \cdot + NO \cdot \longrightarrow RCH_2O \cdot + NO_2 \cdot$$

The NO_2 formed can absorb radiation to produce more ozone. Oxidation of hydrocarbons by ozone may result in polymerisation and the formation of aerosols which create the brown haze of photochemical smog [see § 3.8].

Ozone levels vary from area to area and also throughout the day.

Ozone levels in the troposphere vary from area to area. In towns the ozone level varies throughout the day [see Figure 3.12A]. There is seasonal variation in ozone levels with higher ozone concentrations during the summer months because of the higher intensity and longer duration of sunlight. The residence time of ozone in the troposphere is about 60 hours.

3.12.1 EFFECTS OF OZONE

HUMAN HEALTH

At 100–300 ppb, ozone causes breathing difficulties. This level is exceeded for considerable periods of time in urban areas. Symptoms are reduced by not taking physical exercise. In the USA a stage 1 alert is called at 200 ppb about 150 times a year to tell schools to stop physical activities. Stage 2 alerts at 300 ppb are infrequent. No stage 3 alert at 400 ppb has been called since the scale was devised in 1977.

Ozone affects human health, causing breathing difficulties ...

Some of the compounds present in polluted air have a **synergistic effect**; that is, their combined effect is greater than the sum of their separate effects. It has been found that ozone and sulphur dioxide have a synergistic effect. Some of the compounds present in smog react with ozone to form more dangerous substances which are carcinogenic or mutagenic (causing genetic changes) or both.

PLANTS

Plants are very sensitive to raised ozone levels as low as 40 ppb. It is difficult to separate the effects of ozone from the effects of other pollutants such as acid rain [§ 3.6], but it is believed that ozone is the air pollutant that causes most damage to vegetation. The Black Forest in Germany suffers badly, with 50% of the trees showing the effects of pollution. Ozone levels in the forest have been measured and found to exceed levels which tests have shown to have damaged vegetation.

... damages vegetation, reducing crop yields and causing blemishes ...

Damage to agricultural crops is considerable, both in reduced yields and in unsightly blemishes which reduce the market value of crops. Los Angeles has the climate for growing tobacco and citrus fruits but these crops cannot be grown profitably because of such blemishes.

MATERIALS

Substances with double bonds can be attacked by ozone. Ozone causes rubber to crack, with damage to car tyres. Plastics lose their pliability.

... damages materials ...
... and contributes to the greenhouse effect.

GREENHOUSE EFFECT

Ozone absorbs infrared radiation and can therefore contribute to the enhanced greenhouse effect [see § 3.13].

3.12.2 CONTROL OF OZONE

Ozone is a secondary pollutant. Measures to reduce ozone formation depend on reducing vehicle traffic.

Ozone is a secondary pollutant. Measures to reduce ozone concentrations depend on reducing the emission of primary pollutants. The amounts of carbon monoxide, hydrocarbons and oxides of nitrogen emitted by motor vehicles is half of the total given off by burning fossil fuels. A number of cities have tried to reduce the amount of traffic. Osaka in Japan and Munich in Germany have improved public transport to reduce the number of private vehicles on the roads. Athens in Greece and Florence and Rome in Italy have schemes to allow individual cars to use the city only on alternate days. Catalytic converters are playing an important role [see § 3.9].

CHECKPOINT 3.12

1. Refer to Figure 3.12A showing ozone levels at different times of day.

(*a*) What causes the early morning rise in the NO level?

(*b*) Why does the rise in NO_2 follow the rise in the NO level?

(*c*) Why is there a second peak in NO and in NO_2 levels?

(*d*) Why is there one peak a day in ozone level?

3.13 GLOBAL WARMING

Global warming may be taking place through an increase in the concentrations of greenhouse gases.

The **greenhouse effect** is a natural atmospheric phenomenon, not a result of pollution and was therefore discussed in § 2.9. Carbon dioxide, methane, oxides of nitrogen, ozone and other gases are **greenhouse gases**. The atmosphere has mechanisms for keeping the concentrations of these gases at safe levels. Sometimes human intervention in the natural order of things raises the levels of these naturally occurring gases to high levels at which they become pollutants. If the concentration of greenhouse gases falls, the Earth's temperature will fall. There are indications that the concentrations of greenhouses gases are rising, and **global warming** is taking place. The evidence was discussed in § 2.9.

3.14 RADON

Radon is a naturally occurring radioactive gas.

Radon is a naturally occurring radioactive gas. Radon-222 is formed from the decay of uranium-238. It has a half-life of 3.8 days, decaying through a number of radioactive intermediates to the stable isotope lead-210. Radon-220 is called thoron; it has a half-life of 55 seconds, and radon-219 is called actinon. It is the decay of radon-222 that is known to cause lung cancer. It decays by α-emission and the α-radiation is released into the respiratory system. It has been established that 2500 deaths a year in the UK are due to lung cancer caused by radon.

A link has now been established between radon and leukaemia (cancer of the blood cells). A team of research workers in the University of Bristol have found a link between exposure to radon and the incidence of myeloid leukaemia (bone marrow leukaemia). The level of radon in different geological areas in Canada shows a significant correlation with the incidence of myeloid leukaemia. Radon tends to concentrate in fat cells within bone marrow. α-radiation from the cells may damage blood cells in surrounding tissues. The Bristol team has performed a calculation of the ratio of the occurrence of leukaemia due to radon to the occurrence due to other types of natural radiation. They deduce that α-radiation is more likely to cause damage than previously estimated. Previously it had been accepted that α-radiation is 20 times as damaging to tissues as γ-radiation. The new estimate is a factor of 180.

Leukaemia 'clusters' (areas where the incidence of leukaemia is higher than average) have been observed in Britain. Some of these may be due to radon. α-particles cause more damage to tissues than β-particles because they have twice the charge and are far more massive and are moving relatively slowly so that they are more likely to hit other atoms. α-particles can change DNA chemically. They can ionise water molecules in the body to form free radicals which are highly reactive species able to damage cells. Outside the body, α-particles are not very dangerous because of their short range. They cannot penetrate tissues as β-rays and γ-rays can. If the radioactivity arises inside the body, the α-particles are in position to exercise their damaging effect at short range.

There is a link between radon and leukaemia.

Uranium minerals are common in granite, in metal-bearing rocks, in some hydrocarbon deposits and in some sedimentary rocks containing phosphates. If radon is formed in the atmosphere, it disperses. If it is formed inside buildings it can accumulate. The accumulation can damage the lungs and increase the risk of lung cancer. In 1987 the Government set a level of radon in homes, but in 1990 further research led the Government to halve its original level down to 200 becquerels per cubic metre. (1 becquerel = 1 disintegration per second.) The National Radiological Protection Board calculates that radon and thoron contribute 51% of the average annual radiation dose which the British population receives. The action taken involves methods of reducing the entry of radon into homes and methods of removing it once it has entered.

In some parts of the UK radon levels are high. It is formed by the decay of uranium, which is present in granite and other rocks.

When granite is used as a building material, radon levels inside buildings are high.

Radon enters houses through cracks in floors and walls, joints, gaps around service pipes and cavities in walls. The effect is greatest when there is a big temperature difference between inside the house and outside. Radon enters houses from the building materials themselves, especially in Devon and Cornwall where granite is used. Maintaining air circulation removes radon; this can be done by opening windows and by installing extractor fans.

There are methods of reducing radon levels in buildings.

Radon levels are measured by means of an α-track detector. This is an inexpensive plastic detector which can be left in a building for some weeks, say 4 weeks, while α-particles make tracks in the plastic. After etching in $6 \, mol \, dm^{-3}$ aqueous sodium hydroxide, the tracks can be seen in a microscope at a magnification of ×100 to ×200. From the number of α-particle tracks and the area of the field of view the radioactivity can be calculated.

Radon levels are measured by means of an α-track detector.

CHECKPOINT 3.14

1. Radon levels are found to vary
(*a*) during the day, being highest at night

(*b*) with the seasons, being highest in the winter.
Suggest explanations for these observations.

3.15 ACCIDENTAL POLLUTION

The chemical industry has a good record as far as accidents are concerned, but when an accident does happen, it can be a disaster. The Bhopal disaster illustrates the point. Even after ten years, people are still not sure exactly what caused the accident.

In December 1984 a Union Carbide plant in Bhopal, India, released a cloud of poisonous gas which killed 2500 people and injured 200 000 more.

In December 1984 a cloud of poisonous gas swept through the Indian city of Bhopal, killing 2500 people and injuring 200 000 more. The gas came from a Union Carbide factory making a pesticide called Carbaryl® or Sevin®. On the night of the accident, Union Carbide did not inform the authorities or tell the hospitals what to do. As victims crowded into the hospital, the company medical officer told doctors that the gas was methyl isocyanate, that it was non-poisonous, that it was like tear gas, making the eyes water, and that applying water would bring relief. Even 15 days later, when thousands had died, the works manager was still defending this statement and saying that he knew of no fatalities from methyl isocyanate. The doctors trying to treat the victims did not know what had poisoned them.

In the manufacture of the pesticide the following reactions take place:

$$COCl_2(g) \quad + \quad CH_3NH_2(g) \longrightarrow CH_3NCO(l) + 2HCl(g)$$

| Carbon dichloride oxide (phosgene) | Methylamine | Methyl isocyanate |

        CH_3NCO + $\longrightarrow$

Methyl isocyanate   1-Naphthol   Carbaryl

The escaping gas was said by Union Carbide to be methyl isocyanate and to be non-poisonous. Doctors thought that it might have contained phosgene because it killed foliage. Some of the symptoms resembled cyanide poisoning. At 4000 °C, methyl isocyanate decomposes to give hydrogen cyanide. When victims were treated with sodium thiosulphate, their symptoms were alleviated and they excreted thiocyanate faster in their urine. Union Carbide gave no help or advice to the doctors treating the injured.

Phosgene is a poisonous gas that has been used in chemical warfare, methylamine is a non-toxic gas with a fishy smell, and the effects of methyl isocyanate were unknown. At first doctors in Bhopal thought the gas was phosgene (T_b 8 °C) because phosgene would be more likely to vaporise on a cool evening than methyl isocyanate (T_b 39 °C) and also because the gas had damaged plants for miles around the factory, just as phosgene does. The effects of phosgene are known because it was used in the First World War. It causes mild irritation at first, until it is hydrolysed in the body to carbon monoxide and hydrogen chloride which cause pulmonary oedema: lungs swollen up with water. Phosgene kills the victim within two days or more, but in Bhopal many people were killed immediately. As days went by, it seemed that a mixture of gases was at work, one acting quickly and one producing symptoms after two to three days.

Local investigators began to think that methyl isocyanate had broken down at the raised temperature of the storage tank to form hydrogen cyanide; this would explain the speed with which some of the victims died. There are research papers which describe decomposition of methyl isocyanate at 400 °C. The Union Carbide report on the accident did not identify the mixture of gases that escaped. Union Carbide maintained that methyl isocyanate cannot lead to permanent damage or long-term effects. The victims suffered in the uncertainty about what gases they had inhaled. Sodium thiosulphate is used as a treatment for cyanide poisoning; it causes excretion of cyanide as thiocyanate, CNS^-. Union Carbide said that it was not necessary or advisable to use sodium thiosulphate to treat the survivors. The blood of victims was dark cherry-red, showing that some poison was blocking the use of oxygen; this would fit in with cyanide poisoning. The possibility that methyl isocyanate breaks down in the body to cyanide was suggested. When survivors were treated with thiosulphate, their use of oxygen improved. Two months after the disaster, the Indian Council for Medical Research started treating survivors with thiosulphate, and found that it alleviated their symptoms and increased the excretion of thiocyanate in their urine. Even after this, very few people received thiosulphate treatment.

The death toll continues to rise as more people die of lung diseases. Thousands of people suffered impairment of their eyesight.

The number of casualties is still increasing as people continue to die from lung diseases. Thousands of survivors have been unable to work because of their failing lungs. For some months after the accident there was an increase in women's diseases and in still births and in the number of babies born with defects. The eyes of over 70% of the population were affected, and at first it was thought that thousands would go blind. Although thousands of people now have seriously impaired eyesight and eye irritation, fortunately blindness has not been one of the permanent injuries inflicted by the gas.

How did the accident happen? At 11 p.m. on 2 December 1986 the temperature of a tank containing 3840 gallons of methyl isocyanate rose to 38 °C. As the temperature rose, the liquid vaporised and the pressure in the tank increased. Valves are fitted to the storage tanks to open automatically if the pressure rises and allow gas to escape into 'scrubbers', filled with sodium hydroxide which converts the gas into harmless products:

$$CH_3NCO(g) + 2OH^-(aq) \longrightarrow CH_3NH_2(g) + CO_3{}^{2-}(aq)$$

Methyl isocyanate   Methylamine   Carbonate ion

A rise in temperature of a storage tank holding methyl isocyanate started the chain of events which led to the accident. Valves failed to open to allow gas to escape into 'scrubbers' and release the pressure. The refrigeration unit did not work, one scrubber was out of action, and the flare tower failed to ignite and burn the escaping gas.

The automatic relief valves were not working, and the emergency valves did not work either. The refrigeration plant which might have cooled the storage tank was not working. Two men were sent with hosepipes to cool the tank, but as the pressure inside the tank continued to rise, they ran away. At 1 a.m. a valve ruptured and methyl isocyanate gas surged towards the two scrubber tanks. Only one scrubber was working: the other had been shut down for repairs. The one scrubber could not cope, and gas poured out of it. There was a last line of defence, a flare tower which was designed to burn off escaping gas; this flare failed to ignite. The nightshift fled, while a lone supervisor struggled for 45 minutes to stem the flow of gas. All 3840 gallons of methyl isocyanate escaped. Its dense vapour rolled across the ground to the shanty towns just across the road. It left a trail of dead, blinded and choking people.

There must have been a reason for the liquid methyl isocyanate to have reached 38 °C on a cool evening. A chemical reaction must have been taking place in the storage tank. Methyl isocyanate may have reacted with impurities present in it or in the nitrogen that was fed into the tank. There must have been serious failures in the purification and detection systems for this to happen — unless warning from the detection systems were ignored.

Union Carbide gave little assistance to the injured. The local emergency services were unable to cope with a disaster of this magnitude.

Union Carbide had never instructed the population of Bhopal what to do if a leak occurred. If people had known that the best thing to do was to stay indoors and cover their faces with wet towels, hundreds of lives could have been saved. The plant's alarm system did not go off until three hours after the accident. Local government proved to be incapable of tackling a major emergency. India does not have the kind of emergency services that exist in advanced countries. It is interesting that in the Union carbide plant in France there is a computerised system which is sensitive enough to detect 0.3 ppm of methyl isocyanate and to switch on sprinklers if the temperature of the liquid rises. At the Union Carbide plant in West Germany there is a mobile chemical emergency unit which can smother a leaking tank with foam and suck up gallons of liquid or gas, yet no one is allowed to live within 2 km of the plant.

CHECKPOINT 3.15

1. (*a*) Why do people in developing countries live near chemical factories?

(*b*) What can be done to improve the chances of the local people if there is an accident?

(*c*) Why was Union Carbide able to get away with less stringent safety precautions in India than in France and Germany?

2. (*a*) What precautions could Union Carbide have taken to avoid the accident that happened at Bhopal?

(*b*) After the accident had happened, what assistance could Union Carbide have given to the injured?

QUESTIONS ON CHAPTER 3

1. (*a*) Describe briefly what is meant by the greenhouse effect in the atmosphere. Name three gases which contribute to the greenhouse effect.

(*b*) How does the presence of particulate pollutants in air counteract the greenhouse effect?

(*c*) What will happen to the level of the oceans if there is an enhancement of the greenhouse effect?

2. (*a*) The concentrations of NO and NO_2 in polluted air are not toxic. Why then are these oxides considered to be so undesirable in air?

(*b*) Why are particulates such dangerous pollutants even though they may be chemically inert solids?

3. The composition of the pollutants emitted by a car exhaust depends on the tuning of the engine and also on the mode of operation of the engine.

	Concentration of pollutant/ppm		
	Carbon monoxide	*Hydrocarbons*	*Nitrogen oxides*
Mode A: car with engine idling	50 000	750	15
Mode B: car accelerating	20 000	100	3000

(*a*) Suggest why there is a difference between the concentrations in Modes A and B of

(i) carbon monoxide

(ii) hydrocarbons

(iii) nitrogen oxides.

(*b*) Car exhaust gases contribute to the formation of photochemical smog. Which component is responsible for the yellow colour of photochemical smog?

(*c*) What conditions are necessary for the formation of photochemical smog?

(*d*) How does a catalytic converter deal with the exhaust gases of a car?

4. (*a*) State three uses of CFCs. Explain how the properties of CFCs make them suited to these uses.

(*b*) Explain the contribution which CFCs make to

(i) the thinning of the ozone layer

(ii) the greenhouse effect.

5. (*a*) What is a temperature inversion?

(*b*) Why does Los Angeles have frequent temperature inversions?

(*c*) Mention two harmful consequences which follow from temperature inversions and explain why.

(*d*) Why are the consequences most apparent in the middle of the day rather than in the morning rush hour?

6. (*a*) What essential function does the ozone layer perform for human beings?

(*b*) How is ozone produced and destroyed in the atmosphere?

(*c*) State three ways in which humans release CFCs into the atmosphere.

(*d*) Explain, with equations, how CFCs destroy atmospheric ozone.

(*e*) How are levels of ozone and CFCs in the atmosphere measured?

7. (*a*) What is the main species responsible for oxidising NO to NO_2 in a smoggy atmosphere?

(*b*) At what point in the smog-producing chain reaction is PAN formed?

(*c*) When hydrocarbons are non-toxic, why are they regarded as damaging environmental pollutants?

8. (*a*) When NO_2 absorbs a photon of light, which one of the following is *least* likely to be produced?

O, NO, NO_2*, N

(*b*) Which one of the following species is most likely to be present in a reducing smog?

O, O_3, SO_2, PAN, PBN

(*c*) Which one of the following reaches its peak last on a smog-forming day?

NO, oxidants, hydrocarbons, NO_2

9. (*a*) Why is acid rain classified as a secondary pollutant?

(*b*) How does modern transport contribute to the formation of acid rain?

10. (*a*) How does the extremely low temperature of stratospheric clouds in Antarctica contribute to the ozone 'hole'?

(*b*) Explain why people are so concerned about a decrease in ozone in the stratosphere when ozone is a toxic substance.

4

THE HYDROSPHERE

4.1 THE WATER CYCLE

Water circulates from land and ocean to atmosphere and back to land and ocean in the water cycle...

... which is driven by the absorption of energy from the Sun ...

Water covers 70% of the Earth's surface. There is constant movement of water from land and oceans to the atmosphere and a return of water from the atmosphere to land and oceans by precipitation. The sequence of events is called the **water cycle** [see Figure 4.1A]. The water cycle is driven by the absorption of energy from the Sun. Of the water that evaporates from the oceans and land, 83% comes from the oceans. Of the water that falls as rain and snow, 76% falls on the oceans. There is a net transfer of water from oceans to land. The extra water added to the land returns to oceans in rivers and by seepage of groundwater into oceans. The average **residence time** for water in the atmosphere is 11 days.

$$\text{Residence time} = \frac{\text{Amount of substance in reservoir}}{\text{Rate of addition or removal of substance}}$$

The residence time of water on land varies from a few days for surface run-off to thousands of years for groundwater and hundreds of thousands of years for ice caps. The residence time of water in the oceans is about 4000 years.

FIGURE 4.1A
The Water Cycle (Figures show the mass of water in gigatonnes per day)

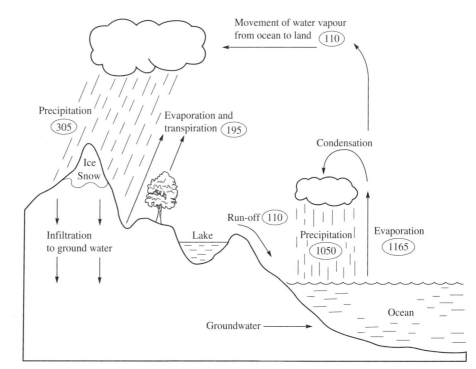

... and transfers heat from one region to another.

Water evaporates most rapidly where the temperature is highest and absorbs the enthalpy of vaporisation in doing so. Heat is released when water vapour condenses. The movement of water vapour therefore transfers heat from one region to another and reduces the temperature difference between regions.

Rainwater becomes acidic when it dissolves pollutants out of the air.

Rain dissolves soluble substances present in the atmosphere: oxygen, nitrogen, carbon dioxide, oxides of sulphur and oxides of nitrogen, and is therefore naturally weakly acidic. In areas where industrial processes and the combustion of fossil fuels send large quantities of sulphur oxides and nitrogen oxides into the atmosphere the rain may be 10 to 100 times more acidic than normal [see Acid Rain, § 3.6]. As acidic rainwater runs over the land and through the surface it dissolves soluble substances from soil and rocks. The water that runs off the land into rivers therefore contains more dissolved substances, e.g. calcium ion, than rainwater.

CHECKPOINT 4.1

1. Most of the rainfall falls on oceans. How is some of this water transferred to land?

2. How does the water cycle transfer heat from one region to another?

3. Why is rainwater weakly acidic? Under what conditions can it become more strongly acidic?

4.2 THE WATER MOLECULE

The shape of the water molecule – a bent line – is illustrated.

Water consists of H_2O molecules in which the angle between the bonds is 104.5°. The oxygen atom in H_2O has two unshared electrons. These two 'lone pairs' of electrons and the two bonding pairs of electrons which form the H—O bonds point to the vertices of a tetrahedron with the oxygen atom at the centre [Figure 4.2A]. In a regular tetrahedron the angle between the bonds is 109.5°. Lone pairs are closer to the nucleus than bonding pairs and there is more repulsion between them than between bonding pairs, so the angles *a* and *b* in Figure 4.2A are greater than 109.5° and the angle of the H—O—H bonds is 104.5° [see *ALC*, § 5.1.3].

FIGURE 4.2A
The Shape of the Water
Molecule

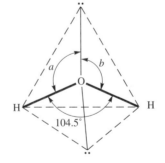

4.3 HYDROGEN BONDING IN WATER

Oxygen is more electronegative than hydrogen and each bond in H_2O is therefore polar:

As a result the molecule has a dipole moment. There is an attraction between the hydrogen atom in one molecule and the oxygen atom in a neighbouring molecule:

The H_2O molecule has a dipole moment.

$$H^{\delta+}\!\!-\!\!O^{\delta-} \cdots\cdots H^{\delta+}\!\!-\!\!O^{\delta-}$$
$$H^{\delta+} \qquad\qquad H^{\delta+}$$

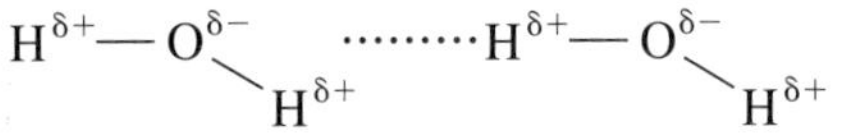

This attraction is a **hydrogen bond**; it has a strength of $\frac{1}{10}$ to $\frac{1}{20}$ of that of a covalent bond [see *ALC*, §4.7.3].

FIGURE 4.3A
Hydrogen Bonding in Water

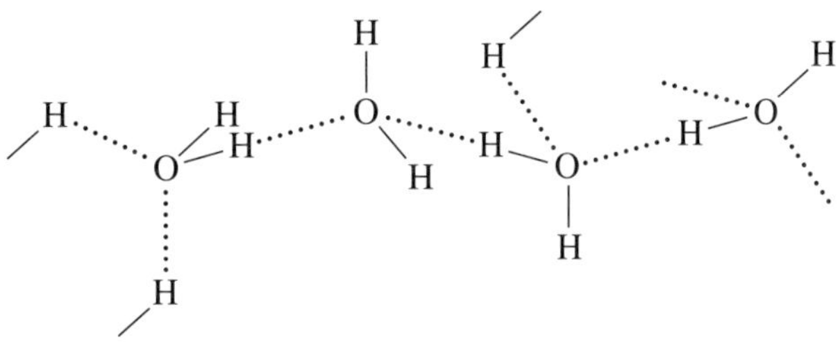

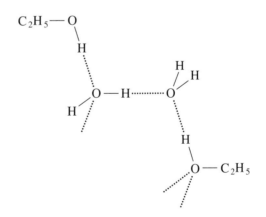

The attraction between a hydrogen atom in one molecule and an oxygen atom in a neighbouring molecule is called a hydrogen bond.

The melting and boiling temperatures of water are much higher than those of the other hydrides in Group 6 , H_2S, H_2Se, H_2Te [see *ALC*, §4.7.3, and Figures 4.33, 4.34]. The reason is that the hydrogen bonds between water molecules are stronger than the intermolecular forces in the other hydrides.

4.4 SOLUBILITY OF ORGANIC COMPOUNDS

Water molecules are associated by hydrogen bonding.

Water is a hydrogen-bonded association of H_2O molecules. A substance such as ethanol, C_2H_5OH, will dissolve in water because molecules of ethanol can displace water molecules in the association. New hydrogen bonds form between molecules of ethanol and water [see Figure 4.4A]. Compounds such as halogenoalkanes, e.g. chloroethane C_2H_5Cl, do not form hydrogen bonds with water and are only slightly soluble.

FIGURE 4.4A
Hydrogen Bonds between Ethanol and Water

Molecules of soluble organic substances are able to displace water molecules in the association.

$$C_2H_5\!\!-\!\!O$$

4.5 SOLUBILITY OF INORGANIC COMPOUNDS

The polar nature of water molecules explains why water is a good solvent for **ionic compounds**. Energy is needed to break the bonds between ions in a crystal and allow the crystal to dissolve. When an ionic solid dissolves in water, the ions become **hydrated**: bonded to water molecules.

FIGURE 4.5A
Water Molecules
Surrounding (a) a Cation
and (b) an Anion

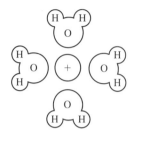

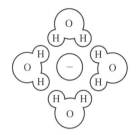

A hydrated cation.
Note that the δ− oxygen atoms of H_2O
molecules point towards the cation.

A hydrated anion.
Note that the δ+ hydrogen atoms of H_2O
molecules point towards the anion.

Ionic compounds dissolve when the energy required to dissociate the crystal structure is less than the energy given out when the ions are hydrated to form aqua ions.

As these forces of attraction come into play, energy is given out. This compensates for the energy required to break up the crystal structure.

Hydrated ions are called **aqua ions**. Many are surrounded by six water molecules [see Figure 4.5B].

FIGURE 4.5B
The Aquasodium Ion,
$[Na(H_2O)_6]^+$ or $Na^+(aq)$
for Short

A saturated solution can dissolve no more solute at that temperature.

When there are so many hydrated ions in a solution that no more can be accepted, the solution is said to be **saturated**. A **saturated solution** is defined as a solution that contains the maximum concentration of dissolved solute at that temperature. The solubility of most solids in aqueous solution increases with a rise in temperature.

During weathering of rocks, an ionic structure is broken down and hydrated ions are formed. Firstly, energy must be supplied to dissociate the lattice. The **lattice enthalpy** is the energy taken in when 1 mole of an ionic compound is formed from its gaseous ions. It has a negative value, e.g.

The standard lattice enthalpy, $\Delta H_{lattice}^{\ominus}$ ...

$$Na^+(g) + Cl^-(g) \longrightarrow NaCl(s); \quad \Delta H_{lattice}^{\ominus} = -780 \, kJ \, mol^{-1}$$

... and the standard enthalpy change of hydration, $\Delta H_{hydr}^{\ominus}$ ...

This means that energy is given out when the lattice is formed, and the energy that must be supplied to dissociate the lattice is $-\Delta H_{lattice}^{\ominus}$. Secondly, after a lattice has been dissociated, the ions dissolve to form hydrated ions. Bonds are formed and energy is given out. The **enthalpy change of hydration** is the enthalpy taken in when 1 mole of a gaseous substance dissolves in an unlimited amount of solvent.

... determine the standard enthalpy change of solution, $\Delta H_{soln}^{\ominus}$ ...

$$Na^+(g) + Cl^-(g) + aq \longrightarrow Na^+(aq) + Cl^-(aq); \quad \Delta H_{hyd}^{\ominus} = -783 \, kJ \, mol^{-1}$$

When a mineral is weathered, the **enthalpy change of solution** is

$$\ldots \Delta H^{\ominus}_{soln} = -\Delta H^{\ominus}_{lattice} + \Delta H^{\ominus}_{hydr}$$

$$NaCl(s) + aq \longrightarrow Na^+(aq) + Cl^-(aq); \quad \Delta H^{\ominus}_{soln}$$

$$\Delta H^{\ominus}_{soln} = -\Delta H^{\ominus}_{lattice} + \Delta H^{\ominus}_{hyd} = +780 - 783 = -3 \, kJ \, mol^{-1}$$

Since $\Delta H^{\ominus}_{soln}$ is exothermic, dissolution is favoured on enthalpy grounds. Enthalpy of hydration is greatest for small ions, e.g. Li^+, and for highly charged ions, e.g. Al^{3+}.

Dissolution is favoured on entropy grounds.

Dissolution is also favoured by the associated increase in **entropy** (disorder). When ions break out of their positions in a crystal structure and spread through a solution there is an increase in entropy.

4.6 SOLUBILITY PRODUCT

Many salts which appear to be insoluble in water do in fact dissolve to a small extent. For such salts a **solubility product** can be quoted. For silver chloride, $AgCl$, the solubility product K_{sp} is given by

For sparingly soluble salts, a solubility product is quoted . . .

$$K_{sp} = [Ag^+(aq)] \, [Cl^-(aq)]$$

When lead(II) fluoride dissolves, one formula unit of the compound, PbF_2, gives $Pb^{2+}(aq) + 2F^-(aq)$. The solubility product is given by

$$K_{sp} = [Pb^{2+}(aq)] \, [F^-(aq)]^2$$

. . . the product of the concentrations of the aqueous ions . . . . . . raised to the appropriate powers.

The solubility product of a salt is defined as the product of the concentrations of its aqueous ions raised to the appropriate powers. The solubility product is related to the **solubility** of the salt. For example, given the solubility of silver chloride $= 1.1 \times 10^{-5} \, mol \, dm^{-3}$, the solubility product can be found.

$$[Ag^+(aq)] = [Cl^-(aq)] = 1.1 \times 10^{-5} \, mol \, dm^{-3}$$

$$K_{sp} = [Ag^+(aq)] \, [Cl^-(aq)] = (1.1 \times 10^{-5} \, mol \, dm^{-3})^2 = 1.2 \times 10^{-10} \, mol^2 \, dm^{-6}$$

4.7 COMPLEXING AGENTS

Many metals are not attacked by water because they form protective coatings of oxides, carbonates and other insoluble compounds. A complexing agent in contact with such metals can dissolve the protective coating to form soluble complex ions and expose the metal beneath to corrosion. **Complex ions** are formed by the combination of an ion with a neutral molecule or an oppositely charged ion. Coordinate bonding is involved. Examples of the formation of soluble complex ions are

The solubility of sparingly soluble compounds may be increased by reaction with complexing agents to form soluble complex ions.

$$Zn^{2+}(aq) + 4NH_3(aq) \longrightarrow [Zn(NH_3)_4^{2+}](aq)$$

$$Cu^{2+}(aq) + 4Cl^-(aq) \longrightarrow [CuCl_4^{2-}](aq)$$

$$Cd^{2+}(aq) + 4CN^-(aq) \longrightarrow [Cd(CN)_4^{2-}](aq)$$

Solutions of complexing agents are employed to clean metal surfaces in metal plating operations. The effluent may be discharged into metal water pipes. There it can convert insoluble compounds of lead, chromium, zinc and other metals into soluble complexes, thus increasing the pollution by heavy metals [see § 6.9].

CHECKPOINT 4.7

1. What is meant by the statement that hydrogen bonds are present between water molecules?

2. Explain the following statements.

(*a*) Oil and water do not mix.

(*b*) Whisky and water do mix.

(*c*) Water is a good solvent for ionic compounds.

(*d*) Many classes of organic compounds dissolve in water.

3. When inorganic compounds dissolve in water, energy must be supplied to separate the ions from the crystal structure. Why is this endothermic reaction able to occur?

4. Write an expression for the solubility product of
(*a*) $CaSO_4$ (*b*) Ag_2SO_4 (*c*) $Fe(OH)_3$.

5. Calculate the solubility of calcium carbonate
(*a*) in $mol\,dm^{-3}$ and (*b*) in $g\,dm^{-3}$, given the solubility product = $5.0 \times 10^{-9}\,mol^2\,dm^{-6}$ at 25 °C.

6. The solubility of lead(II) carbonate at 25 °C is $2.5 \times 10^{-7}\,mol\,dm^{-3}$. Calculate the solubility product at this temperature.

7. Copper becomes weathered in air to acquire a surface layer of copper carbonate hydroxide, $Cu(OH)_2 \cdot CuCO_3$. Explain how this surface layer is removed by

(*a*) aqueous ammonia

(*b*) aqueous sodium chloride

(*c*) aqueous potassium cyanide

4.8 ICE

Ice is a hydrogen-bonded structure of water molecules, which spaces the molecules further apart than they are in water.

Water molecules are associated by hydrogen bonding [Figure 4.3A]. In ice the arrangement of water molecules is similar, but the regularity extends through the whole structure [see Figure 4.8A]. The structure spaces the molecules further apart than they are in liquid water. This is why when water freezes it expands by 9% and why ice is less dense than water at 0 °C. When ice melts about 15% of the hydrogen bonds are broken, and the open structure collapses, thus increasing the density. As more heat is supplied, the kinetic energy of the molecules increases so they vibrate more and therefore occupy a larger volume, thus decreasing the density. The two changes produce opposite effects, and water has a maximum density at 4 °C.

FIGURE 4.8A
Hydrogen Bonding in Ice
Notes   Each H_2O molecule uses both H atoms to form hydrogen bonds and is also bonded to two other H_2O molecules through their H atoms. The arrangement of bonds about the O atoms is tetrahedral.

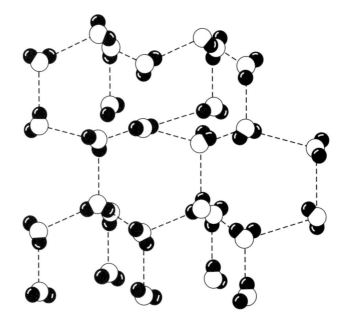

This is why ice is less dense than water at the same temperature.

The fact that water reaches its maximum density at 4 °C explains why ponds and lakes freeze from the surface downwards. As it cools below 4 °C, the water at the surface becomes less dense and therefore stays on top of the slightly warmer water until it freezes. Additional hydrogen bonds form in freezing with the release of energy which

Water has maximum density at 4 °C, so lakes freeze from the top downwards . . .

raises the temperature of the water below the ice. The layer of ice on the surface helps to insulate the water below from further heat loss. Aquatic life can continue below the surface unless the water is very shallow and unless the air temperature remains below 0 °C all year round. Fish and plants survive under the ice in Canadian lakes and rivers for months.

. . . and ice is a weathering agent.

The **weathering** of rocks and soils is assisted by the expansion that occurs when water freezes. Water trapped in cracks builds up large forces when it freezes and attempts to expand. These forces can cause the rocks to shatter.

4.9 HEAT CAPACITY OF WATER

The quantities of heat required for changes in water are:

Water has a high specific heat capacity . . .

- to convert ice into water at $0 \,°C = 320 \, J \, g^{-1}$
- to vaporise water $= 2260 \, J \, g^{-1}$
- to raise the temperature of liquid water $= 4.2 \, J \, K^{-1} \, g^{-1}$

The quantities of heat required for these changes are much greater than those for other substances, and water is therefore a good medium for controlling changes in temperature. In living organisms chemical reactions in the cells often generate large quantities of heat; these can be absorbed by water with only a small rise in temperature. If solar energy is providing too much energy for an organism, the vaporisation of small amounts of water by transpiration in plants and perspiration in animals provides an effective cooling system. The vaporisation of 1 g of water absorbs 500 times as much heat as raising the temperature of 1 g of water by 1 K.

The **specific heat capacity** of a substance is defined by

$$\text{Specific heat capacity} = \frac{\text{Heat required to raise the temperature of substance/J}}{\text{Mass/g} \times \text{Temperature rise/K}}$$

The unit is $J \, K^{-1} \, g^{-1}$, and the value of the specific heat capacity of water is $4.2 \, J \, K^{-1} \, g^{-1}$.

The **heat capacity** of a body is defined by

$$\text{Heat capacity} = \frac{\text{Heat required to raise the temperature of a body/J}}{\text{Temperature rise/K}}$$

. . . so a large body of water has a stabilising effect on the temperature of a region.

The unit is $J \, K^{-1}$. The heat capacity of a large body of water tends to stabilise the temperature of nearby land masses by absorbing heat in summer and releasing it in winter, so reducing seasonal temperature differences.

CHECKPOINT 4.9

1. Water has maximum density at 4 °C.

(*a*) Explain why this is the case.

(*b*) Comment on how this property affects (i) aquatic life, (ii) weathering of rocks.

2. Suggest why the standard enthalpy of vaporisation of water is higher than that of ethanol.

3. Explain how water is able to assist (i) plants, (ii) animals to avoid a rise in temperature.

4. Define the specific heat capacity of water. What is its value? A lake is 2 km long, 1 km wide and 100 m deep. Calculate the heat required to heat it by 1°C (using the specific heat capacity of pure water).

4.10 WATER AS AN ACID AND AS A BASE

Brönsted and Lowry gave the following definition of acids and bases [see *ALC*, § 12.7.1]. An **acid** is a substance that can donate a proton to another substance. A **base** is a substance that can accept a proton from another substance. No substance can act as an acid in solution unless a base is present to accept a proton: the reactions of acids are reactions between acids and bases. Similarly all reactions of bases in solution are acid–base reactions. Examples of acid–base reactions are:

Water can act as an acid – a proton donor – and also as a base – a proton acceptor.

1. $HCl + H_2O \longrightarrow H_3O^+ + Cl^-$

2. $CH_3CO_2H + H_2O \rightleftharpoons H_3O^+ + CH_3CO_2^-$

3. $NH_3 + H_2O \rightleftharpoons NH_4^+ + OH^-$

4. $CH_3CO_2^- + H_2O \rightleftharpoons CH_3CO_2H + OH^-$

Water is an amphoteric or amphiprotic solvent.

In **1** and **2**, water is acting as a proton acceptor, a base; in **3** and **4**, water is acting as a proton donor, an acid. Water is described as an **amphoteric** or **amphiprotic** solvent.

4.11 pK_w

Water is ionised to a slight extent:

$$2H_2O(l) \rightleftharpoons H_3O^+(aq) + OH^-(aq)$$

The product of the concentrations of hydrogen ions and hydroxide ions is called the **ionic product** for water, K_w.

The ionic product of water, $K_w = 1.00 \times 10^{-14}\ mol^2\ dm^{-6}$ at 25°C.

$$K_w = [H_3O^+(aq)]\,[OH^-(aq)]$$

The value of K_w at 25 °C is $1.00 \times 10^{-14}\ mol^2\ dm^{-6}$, and p$K_w$ is 14.

In a neutral solution, the hydrogen ion concentration and the hydroxide ion concentration are both equal to $1.00 \times 10^{-7}\ mol\,dm^{-3}$ at 25 °C, and $pH = -\log[H_3O^+(aq)/mol\,dm^{-3}] = 7$, and $pOH = -\log[OH^-(aq)/mol\,dm^{-3}] = 7$. Acidic solutions have a value of pH $<$ 7, and alkaline solutions have a value of pH $>$ 7.

pK_w = 14, and pH = pOH = 7 in neutral solution

4.12 SALT HYDROLYSIS

Some salts react with water to give acidic or alkaline solutions. These reactions are called salt hydrolysis.

Many salts dissolve in water to give neutral solutions. Some salts, however, react with water to form acidic or alkaline solutions. These reactions are described as **salt hydrolysis**.

Sodium carbonate is the salt of a strong base and a weak acid. The carbonate ion is a base, a proton acceptor. In aqueous sodium carbonate, the reaction which makes the solution alkaline is:

The solution of a salt of a weak acid and a strong base, e.g. Na₂CO₃, is alkaline.

$$CO_3^{2-}(aq) + H_2O(l) \rightleftharpoons OH^-(aq) + HCO_3^-(aq)$$

The CO_3^{2-} ion is a stronger base (proton acceptor) than HCO_3^-; therefore the position of equilibrium lies over to the right-hand side, and the solution is alkaline.

Ammonium chloride is the salt of a weak base and a strong acid. Salt hydrolysis occurs.

The solution of a salt of a strong acid and a weak base, e.g. NH₄Cl, is acidic.

$$NH_4^+(aq) + H_2O(l) \rightleftharpoons NH_3(aq) + H_3O^+(aq)$$

NH_3 is a weak base so the equilibrium lies over to the right-hand side, and the solution is acidic.

The aluminium ion Al^{3+} is small in size and highly charged. In aqueous solution the aluminium ion is hydrated to form the aqua ion, $[Al(H_2O)_6]^{3+}$ [see Figure 4.12A]

FIGURE 4.12A
(a) $[Al(H_2O)_6]^{3+}$
(b) Polarity of O—H bonds

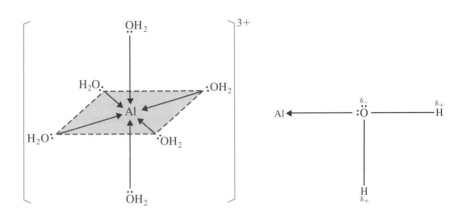

As a lone pair of electrons on the oxygen atom forms a coordinate bond to the aluminium ion, the electrons in the O—H bond move closer to the oxygen atom. The hydrogen atoms have a larger partial positive charge than those in free water molecules. They are therefore more likely to be abstracted by bases, e.g. H_2O, and salt hydrolysis occurs.

$$[Al(H_2O)_6]^{3+}(aq) + H_2O(l) \rightleftharpoons [Al(OH)(H_2O)_5]^{2+}(aq) + H_3O^+(aq)$$

$$[Al(OH)(H_2O)_5]^{2+}(aq) + H_2O(l) \rightleftharpoons [Al(OH)_2(H_2O)_4]^+(aq) + H_3O^+(aq)$$

The small size and high charge of the ions Al^{3+} and Fe^{3+} causes their salts to be hydrolysed, making solutions of their salts acidic.

Hydrogen ions are formed, making aqueous aluminium salts acidic. Salt hydrolysis occurs also in iron(III) salts because of the high charge/size ratio of the ion Fe^{3+}. Aluminium salts and iron(III) salts make soil acidic [see § 7.9]. They are used as coagulants in water treatment [see § 5.4].

CHECKPOINT 4.12

The following are weak acids: hydrocyanic acid HCN, ethanoic acid CH_3CO_2H, hydrogen sulphide H_2S. Ammonia is a weak base.

1. Predict whether the pH of the following solutions will be 7 or > 7 or < 7. Give your reasons.

(*a*) $0.10 \, mol \, dm^{-3}$ ammonium chloride

(*b*) $0.10 \, mol \, dm^{-3}$ potassium cyanide

(*c*) $0.10 \, mol \, dm^{-3}$ sodium ethanoate

2. Explain the following.

(*a*) A solution of sodium sulphide has a nasty smell.

(*b*) It is dangerous to handle an aqueous solution of potassium cyanide.

(*c*) When acidic rainwater passes through limestone rocks, the pH of the water increases.

4.13 HARD WATER

Calcium and magnesium ions form 'scum' with soap.

The presence of calcium and magnesium ions makes water hard. A soap such as sodium octadecanoate, $C_{17}H_{35}CO_2Na$, reacts with calcium and magnesium ions to form an insoluble 'scum' of calcium and magnesium octadecanoate:

$$Ca^{2+}(aq) + 2C_{17}H_{35}CO_2^-(aq) \longrightarrow (C_{17}H_{35}CO_2)_2Ca(s)$$

Detergents lather, even in hard water.

When all the calcium and magnesium ions have been precipitated, then the soap will lather. Detergents do not form scum in hard water. They are the sodium salts of sulphonic acids, and the calcium and magnesium salts are soluble:

$$C_{12}H_{25} \longrightarrow \langle\bigcirc\rangle \longrightarrow SO_3Na$$

Sodium dodecylbenzenesulphonate (a detergent)

4.13.1 TEMPORARY HARDNESS

Temporary hardness is caused by the hydrogencarbonates of calcium and magnesium . . .

Limestone reacts with rainwater that contains dissolved carbon dioxide to form the soluble salt calcium hydrogencarbonate:

$$CaCO_3 + H_2O(l) + CO_2(aq) \underset{\text{boil hard water}}{\overset{\text{limestone and rainwater}}{\rightleftarrows}} Ca(HCO_3)_2(aq)$$

. . . and is removed by boiling.

The reaction is reversed by boiling. When hard water containing calcium hydrogencarbonate is boiled, calcium carbonate is deposited, and the water becomes soft. Hardness caused by hydrogencarbonates is therefore called **temporary hardness**.

4.13.2 PERMANENT HARDNESS

Permanent hardness is caused by calcium and magnesium sulphates . . .

When water trickles over rocks containing calcium or magnesium sulphates, these minerals dissolve. The resulting hard water cannot be softened by boiling, and is described as **permanently hard water**. It can be softened by the following methods, which also remove temporary hardness.

WASHING SODA

. . . and is removed by washing soda . . .

Washing soda, $Na_2CO_3 \cdot 10H_2O$, is added to precipitate calcium and magnesium ions as insoluble carbonates.

ION EXCHANGE

. . . by passing through an ion exchange resin . . .

Hard water can be softened by passing it slowly through a column containing sodium aluminium silicate (Permutit®). An exchange of ions takes place:

$$Ca^{2+}(aq) + 2NaAlSilicate(s) \underset{\text{Regenerating spent Permutit}^{®}}{\overset{\text{Permutit}^{®} \text{ softening hard water}}{\rightleftarrows}} 2Na^+(aq) + Ca(AlSilicate)_2(s)$$

When all the sodium ions in the Permutit® have been replaced by calcium and magnesium ions, the Permutit® must be regenerated. This is done by passing through it a concentrated solution of sodium chloride. After this solution has been washed out of the column, the Permutit® is ready for use. [See *ALC*, § 8.7.6 for the theory of ion exchange.]

CALGON

. . . and by complexing agents.

Calgon® and similar products are the sodium salts of polyphosphate ions. They are able to form insoluble complexes with calcium ions and magnesium ions, releasing sodium ions as they do so.

ESTIMATION

Permanent hardness can be measured by titration.

Permanent hardness can be measured by a titrimetric method. Calcium and magnesium ions form stable complexes with edta. Titration of a sample of water against a standard solution of edta can be used to find the concentration of calcium ion and magnesium ion in solution, that is, the total hardness of the water. [See *ALC*, §17.10.3.]

4.13.3 HEART DISEASE

Hard water areas appear to have lower death rates from heart disease than soft water areas.

Research workers have investigated the relationship between hard water and death from heart disease, which is the leading cause of death in industrial countries. They noted the causes of death in populations of men aged 45–64 in England and Wales in soft-water areas and hard-water areas. They found that in soft-water areas there were 10 000 more deaths from heart disease than in hard-water areas. In European countries, populations living on older rocks (older than 600 million years) have higher death rates from heart disease than those living on younger rocks (younger than 180 million years). The older rocks contain less carbonate and the water is softer. However, there is another factor. The older rocks have lower concentrations of trace elements. In the USA also, higher death rates from heart disease are associated with soft water and with rocks, soils and water deficient in trace elements. The studies appear to show that there is a factor strongly related to water hardness which has an effect on heart disease and strokes. The factor may be hardness (possibly by preventing dissolution of lead and cadmium from water pipes; see §6.9) or it may be beneficial trace elements which are present in hard water.

4.14 PURE WATER

Pure water

Pure water is made by distillation. Many laboratories find it more convenient to purify water by running it through a column containing ion-exchange resins.

Ion exchange resins replace cations by $H^+(aq)$ ...

... and anions by $OH^-(aq)$

Ion exchange is a type of partition of ionic compounds. The **ion exchange resin** is a polymer which contains at intervals polar groups which can remove undesirable cations (or anions) and replace them with other cations (or anions). The water softener Permutit® is sodium aluminium silicate, which replaces calcium and magnesium ions in hard water by sodium ions. To purify water, a resin must replace all the cations and anions present by hydrogen ions and hydroxide ions. A combination of a **cation exchanger** and an **anion exchanger** is needed. Cation exchange resins often contain sulphonic acid groups, $-SO_3^-H^+$, and anion exchangers often contain quaternary ammonium groups, e.g., $-\overset{+}{N}(CH_3)_3OH^-$. If water passes slowly through a **deioniser**, equilibrium is set up at each level between hydrogen ions attached to the resin and hydrogen ions in solution:

$$-SO_3^-H^+(resin) + Na^+(aq) \rightleftharpoons -SO_3^-Na^+(resin) + H^+(aq)$$

As the water moves on, two of the components in the equilibrium ($H^+(aq)$ and $Na^+(aq)$) are removed and a fresh equilibrium is established. Each successive equilibration increases the replacement of metal cations by hydrogen ions. At the same time, hydroxide ions are replacing other anions. If tap water is run slowly through an ion exchange resin, 'deionised' water of high purity can be obtained.

1. (*a*) What is pure water?

(*b*) What is an ion exchange resin?

(*c*) How can an ion exchange resin be employed to give purified water?

2. (*a*) Name a mineral which reacts with rainwater (not simply dissolves) in the formation of hard water. Write an equation for the reaction.

(*b*) Give two methods for softening hard water formed in this way.

4.15 SOLUBILITIES OF GASES

4.15.1 HENRY'S LAW

Dissolved gases are essential to aquatic life. Plants need carbon dioxide and fish need oxygen. The solubilities of gases in liquids are governed by **Henry's Law**. This states: The solubility of a gas in a liquid at a certain temperature is proportional to the partial pressure of the gas in contact with the liquid. The law applies to gases which simply dissolve without reacting with water.

Henry's law states that the solubility of a gas in a liquid is proportional to the partial pressure of the gas at that temperature.

$$[X(aq)] = kP_X$$

where $[X(aq)]$ = aqueous concentration of gas X, P_X = partial pressure of X, k = Henry's Law constant for X at a certain temperature.

4.15.2 SOLUBILITY OF OXYGEN

The solubility of oxygen in water can be calculated.

Dry air is 20.95% by volume oxygen. At 25 °C, the partial pressure of water = 3.20 kPa. Henry's Law constant for oxygen at 25 °C = 1.26×10^{-5} mol dm^{-3} kPa^{-1}. The partial pressure of oxygen , P_{O_2} is given by

$$P_{O_2} = (101.3 - 3.20)\,\text{kPa} \times 0.02095 = 20.5\,\text{kPa}.$$

$$[O_2(aq)] = kP_{O_2} = 1.26 \times 10^{-5}\,\text{mol dm}^{-3}\,\text{kPa}^{-1} \times 20.5\,\text{kPa}$$
$$= 2.58 \times 10^{-4}\,\text{mol dm}^{-3}$$

The solubility of oxygen in water is 8.26 ppm at 25 °C.

This concentration is 8.26×10^{-3} g dm^{-3} or 8.26 ppm (parts per million). Water in equilibrium with air at 25 °C and atmospheric pressure cannot contain a higher level of dissolved oxygen than this.

Decaying organic matter uses up oxygen dissolved in water.

The decay of organic matter consumes dissolved oxygen. We can get an idea of the amount of oxygen used in the oxidation by representing the formula of organic matter as (CH_2O), with molar mass 30 g mol^{-1}.

$$(CH_2O)(aq) + O_2(aq) \longrightarrow CO_2(aq) + H_2O(l)$$

An estimate is that 8 g of decaying organic matter can use all the oxygen dissolved in 1 tonne of water.

Then the 8.26 mg of oxygen dissolved in 1 dm^3 of water is used up in oxidising 7.8 mg of (CH_2O). Only 8 mg of decaying organic matter can remove all the dissolved oxygen in 1 dm^3 of water at 25 °C, and 8 g can deplete 1 tonne of water of dissolved oxygen!

The solubility decreases with rising temperature.

The solubility of a gas decreases with a rise in temperature (because the enthalpy change of solution is negative). The solubility of oxygen halves its value between 0 °C and 35 °C. Aquatic organisms have a higher respiration rate at higher temperatures, and this combined with the decrease in solubility of oxygen may lead to a shortage of dissolved oxygen.

4.15.3 BIOCHEMICAL OXYGEN DEMAND

The biochemical oxygen demand, BOD, of a body of water is the quantity of oxygen used when all the organic matter in the water decays.

The **biochemical oxygen demand** or **biological oxygen demand, BOD**, of a body of water is the quantity of oxygen (in ppm) utilised when the organic matter in the water is degraded biologically. A body of water with a high BOD and no means of replenishing oxygen rapidly cannot sustain organisms that require oxygen. Fish require at least $3\,mg\,dm^{-3}$ of oxygen. Pure water has a BOD of 1 ppm, fairly pure water 3 ppm, water of doubtful purity 5 ppm, and the discharge of water with a BOD of 20 ppm into a stream is polluting.

The BOD of a volume of water can be measured. A water sample is saturated with oxygen so the concentration of dissolved oxygen is known. It is incubated at $25\,°C$ for a known period, usually 5 days, while micro-organisms in the water oxidise organic matter. The oxygen that remains in the water is measured by the Winkler method. The oxygen used up – the BOD – is found by subtraction.

The steps in the Winkler method of finding dissolved oxygen concentration are:

1. A known amount of manganese(II) salt is oxidised to manganese(IV) oxide by oxygen in alkaline solution.

$$2Mn^{2+}(aq) + 4OH^-(aq) + O_2(g) \longrightarrow 2MnO_2(s) + 2H_2O(l)$$

2. Iodine is liberated by the reaction of the manganese(IV) oxide formed with potassium iodide in acidic solution.

It can be found by saturating water with oxygen, waiting for the organic matter to decay, and titrating to measure the oxygen concentration left.

$$MnO_2(s) + 2I^-(aq) + 4H^+(aq) \longrightarrow Mn^{2+}(aq) + I_2(aq) + 2H_2O(l)$$

3. Iodine is titrated against standard sodium thiosulphate solution.

$$I_2(aq) + 2S_2O_3{}^{2-}(aq) \longrightarrow S_4O_6{}^{2-}(aq) + 2I^-(aq)$$

The amount of thiosulphate used in Step 3 gives the amount of oxygen that was used in Step 1.

4.15.4 CHEMICAL OXYGEN DEMAND

The chemical oxygen demand, COD, is the quantity of oxygen needed to oxidise all the organic matter in the water. It can be found by titration.

The determination of BOD is slow and does not include wastes which are oxidised even more slowly. **Chemical oxygen demand, COD**, is the quantity of oxygen (in ppm) required to oxidise all the organic matter in the water. The COD can be found more rapidly than the BOD. A sample of water is treated with a known amount of a powerful oxidising agent, e.g. potassium dichromate(VI). After about 2.5 hours, the excess of oxidant is determined by back-titration against a standard reducing agent. A source of error is the oxidation of chloride ion, which is not a pollutant, to chlorine. This can be prevented by precipitating chloride ion as insoluble mercury(II) chloride by adding aqueous mercury(II) sulphate.

4.15.5 SOLUBILITY OF CARBON DIOXIDE

All natural waters contain carbon dioxide, which comes from the air and from decaying matter.

Carbon dioxide dissolves in water from air where it is present as 0.035% by volume. It is also formed by the decay of organic matter in water. All natural waters therefore contain carbon dioxide. Even rainwater in a completely unpolluted atmosphere contains carbon dioxide. Equilibrium exists between carbon dioxide in the atmosphere and dissolved carbon dioxide:

$$CO_2(g) \rightleftharpoons CO_2(aq)$$

The solubility of carbon dioxide is increased by its reaction with water to form hydrogencarbonate ions, carbonate ions and hydrogen ions:

$$CO_2(aq) + H_2O(l) \rightleftharpoons HCO_3{}^-(aq) + H^+(aq)$$

$$HCO_3{}^-(aq) \rightleftharpoons CO_3{}^{2-}(aq) + H^+(aq)$$

The second dissociation occurs to a very slight extent, and very little $CO_3{}^{2-}(aq)$ is present. In alkaline conditions, $OH^-(aq)$ ions remove $H^+(aq)$ ions from the equilibria:

$$H^+(aq) + OH^-(aq) \rightleftharpoons H_2O(l)$$

It reacts with water to form hydrogencarbonate ions and carbonate ions, which have a buffering effect.

As $H^+(aq)$ ions are removed, the dissociation of $HCO_3{}^-$ ions to form more $H^+(aq)$ ions is promoted. The result is a buffering of the effect of the $OH^-(aq)$ ions.

$$HCO_3{}^-(aq) \rightleftharpoons CO_3{}^{2-}(aq) + H^+(aq)$$

In acidic conditions, the combination of $H^+(aq)$ ions with $CO_3{}^{2-}$ ions and with $HCO_3{}^-$ ions is promoted, thus reducing the acidity.

$$H^+(aq) + CO_3{}^{2-}(aq) \rightleftharpoons HCO_3{}^-(aq)$$

$$H^+(aq) + HCO_3{}^-(aq) \rightleftharpoons CO_2(aq) + H_2O(l)$$

In this way the solution is buffered by hydrogencarbonate ions and carbonate ions.

The relative concentrations of CO_2, $HCO_3{}^-$ and $CO_3{}^{2-}$ depend on the pH.

The way in which dissolved CO_2 is distributed between the different species $CO_2(aq)$, $HCO_3{}^-(aq)$ and $CO_3{}^{2-}(aq)$ therefore depends on the pH. For each mole of dissolved CO_2,

Amount of $CO_2(aq)$ + Amount of $HCO_3{}^-(aq)$ + Amount of $CO_3{}^{2-}(aq)$ = 1 mole

The mole fraction of each species is given by, e.g.

Mole fraction of $CO_2(aq)$ =

$$\frac{\text{Amount of } CO_2(aq)}{\text{Amount of } CO_2(aq) + \text{Amount of } HCO_3{}^-(aq) + \text{Amount of } CO_3{}^{2-}(aq)}$$

Figure 4.15A shows a plot against pH of the mole fractions of $CO_2(aq)$, $HCO_3{}^-(aq)$ and $CO_3{}^{2-}(aq)$. In most waters the hydrogencarbonate ion is the predominant species. In more acidic waters, $CO_2(aq)$ predominates.

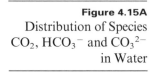

Figure 4.15A
Distribution of Species
CO_2, $HCO_3{}^-$ and $CO_3{}^{2-}$
in Water

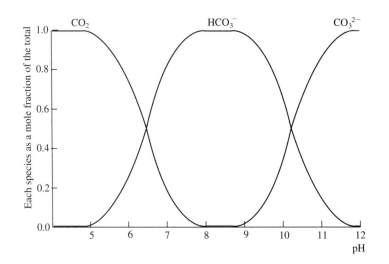

4.15.6 EFFECT OF CALCIUM CARBONATE ON CARBON DIOXIDE EQUILIBRIA

In the presence of calcium carbonate, the equilibria change ...

When solid calcium carbonate is present, it changes the equilibria between dissolved carbon dioxide, hydrogencarbonate ion and carbonate ion. Some calcium carbonate dissolves:

$$CaCO_3(s) \rightleftharpoons Ca^{2+}(aq) + CO_3^{2-}(aq)$$

... as $CO_3{}^{2-}$ ions enter the water from the dissolution of $CaCO_3(s)$ ...

The carbonate ions formed make the solution more alkaline:

$$CO_3^{2-}(aq) + H_2O(l) \rightleftharpoons HCO_3^-(aq) + OH^-(aq)$$

Some calcium carbonate reacts with aqueous carbon dioxide:

$$CaCO_3(s) + CO_2(aq) + H_2O(l) \rightleftharpoons Ca(HCO_3)_2(aq)$$

... and $HCO_3{}^-$ ions enter the water from the reaction of $CaCO_3(s)$ with $CO_2(aq)$.

The hydrogencarbonate ions formed make the solution more alkaline:

$$HCO_3^-(aq) + H_2O(l) \rightleftharpoons CO_2(aq) + H_2O(l) + OH^-(aq)$$

As a result of these basic reactions of carbonate ions and hydrogencarbonate ions, water in equilibrium with atmospheric carbon dioxide and solid calcium carbonate has a pH of 8.3. The value in the absence of calcium carbonate is pH = 5.65. The equilibria are summarised in Figure 4.15B.

The dissolution of limestone is governed by:

(i) the solubility product of calcium carbonate,

(ii) attack by $CO_2(aq)$ to form $Ca(HCO_3)_2(aq)$

FIGURE 4.15B
Carbon Dioxide Equilibria: When $[CO_2(g)]$ in the atmosphere is high the train of events converts CO_2 from $CO_2(g)$ in the atmosphere into $CaCO_3(s)$ as limestone. When $[CO_2(g)]$ in the atmosphere falls, $CaCO_3(s)$ dissolves and the sequence of reactions produces more $CO_2(g)$.

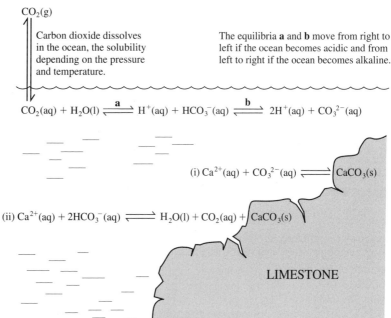

ATMOSPHERE

$CO_2(g)$

Carbon dioxide dissolves in the ocean, the solubility depending on the pressure and temperature.

The equilibria **a** and **b** move from right to left if the ocean becomes acidic and from left to right if the ocean becomes alkaline.

$$CO_2(aq) + H_2O(l) \underset{}{\overset{a}{\rightleftharpoons}} H^+(aq) + HCO_3^-(aq) \underset{}{\overset{b}{\rightleftharpoons}} 2H^+(aq) + CO_3^{2-}(aq)$$

(i) $Ca^{2+}(aq) + CO_3^{2-}(aq) \rightleftharpoons CaCO_3(s)$

(ii) $Ca^{2+}(aq) + 2HCO_3^-(aq) \rightleftharpoons H_2O(l) + CO_2(aq) + CaCO_3(s)$

LIMESTONE

CHECKPOINT 4.15

1. Explain the terms BOD and COD. Briefly outline how each of them can be measured.

2. (*a*) When carbon dioxide dissolves in water, what carbon-containing species are present in solution?

(*b*) How does the solution exert a buffering action?

(*c*) Is the concentration of the following greater at high pH or low pH?

(i) $CO_2(aq)$

(ii) $HCO_3^-(aq)$

(iii) $CO_3^{2-}(aq)$

(d) Which of the species listed in (*c*) predominates in most waters?

(e) When an aqueous solution of CO_2 is in contact with limestone, what changes in equilbria concerning CO_2 occur?

4.16 CALCIUM CARBONATE AND MAGNESIUM CARBONATE

Calcium carbonate and magnesium carbonate have sedimented out from the oceans over geological ages of time.

Before living things populated the oceans, removal of calcium and magnesium from the oceans was governed solely by the solubility products of their compounds. Calcium carbonate has a lower solubility than magnesium carbonate; therefore calcium ion is removed from seawater as a sediment of calcium carbonate in preference to magnesium ion. The residence times in seawater are: Mg^{2+}, 1.5×10^7 years; Ca^{2+}, 1.0×10^6 years. Sediments of calcium carbonate and magnesium carbonate become compressed over geological ages to form limestone, dolomite and other sedimentary rocks [see § 7.2]. The difference in solubility explains why in sedimentary rocks deposits of limestone are much greater than deposits of magnesium carbonate. The production of limestone is favoured by the absence of silicates which would precipitate calcium as calcium silicate. Many limestones have a high degree of purity, with some as high as 98% calcium carbonate.

Calcium carbonate is less soluble than magnesium carbonate. This is why there are more sedimentary rocks containing calcium carbonate.

Limestones have also formed from sediments of the skeletons of marine creatures.

Living things started to populate the oceans 600 million years ago. They provided a different route for the formation of limestones. The majority of marine animals have endoskeletons or exoskeletons that contain calcium. The low solubilities of calcium carbonate, hydroxyapatite, $Ca_3(PO_4)_2 \cdot Ca(OH)_2$ and fluorapatite, $Ca_3(PO_4)_2 \cdot CaF_2$, made them suitable for incorporation into skeletons. When marine organisms died, their skeletons became compressed over geological ages by the weight of material accumulating above them and were slowly transformed into limestone. In time the oceans changed their boundaries and some of these rocks became part of the land [see sedimentary rocks, § 7.2].

Limestone has many industrial uses.

Limestone is used as a source of calcium oxide, CaO (lime or quicklime) which is used in cement, in glass-making, steel-making and sugar-refining. These industries consume millions of tonnes of limestone annually. The UK has plentiful deposits of limestone, but it happens that these deposits are in areas of outstanding natural beauty, and some are in national parks. Conservationists want to preserve the few undeveloped areas in their natural state.

CHECKPOINT 4.16

1. Ionic radii are: Mg^{2+}, 65 pm; Ca^{2+}, 100 pm.

(*a*) Suggest how the difference in size accounts for the higher solubility of magnesium carbonate over calcium carbonate.

(*b*) What is meant by the term 'residence time'?

(*c*) Why is the residence time of $Mg^{2+}(aq)$ longer than that of $Ca^{2+}(aq)$?

2. (*a*) Why do most countries have plentiful deposits of limestone?

(*b*) Mention a few industrial uses of limestone.

(*c*) How does the industrial importance of limestone affect the environment?

FIGURE 4.16A
A Limestone Quarry

QUESTIONS ON CHAPTER 4

1. (*a*) Sketch the structure and shape of the water molecule. Mark the H—O—H bond angle.

(*b*) Describe the intermolecular bonding in water.

(*c*) Why is ethanol soluble in water whereas gasoline is not?

(*d*) Why is sodium chloride soluble in water whereas calcium carbonate is only very slightly soluble?

2. (*a*) Describe the structure of the water molecule and the nature of the bonds.

(*b*) Explain why water is such a good solvent for a variety of solutes.

(*c*) After you have studied Chapter 7, comment on the role of water in physical and chemical weathering of rocks.

3. (*a*) Give an account of the water cycle.

(*b*) Explain why rainwater is naturally acidic.

(*c*) Explain why groundwater tends to be alkaline.

4. Water is unusual in that the density of solid water at 0 °C is lower than the density of liquid water at the same temperature.

(*a*) Explain why this is the case.

(*b*) State what importance this has for (i) aquatic life, (ii) weathering of rocks.

5. (*a*) Some water supplies are described as 'hard'. What practical problems are associated with hard water?

(*b*) Name and give the formula of one compound that causes temporary hardness. Describe how the water may be softened, giving a balanced equation for the reaction.

(*c*) Name and give the formula of a compound that causes permanent hardness. Describe how the water may be softened, giving a balanced equation for the reaction (the method should be different from that in part (*b*)).

6. The concentration of methane in a sample of water is $150 \, cm^3 \, dm^{-3}$ at rtp. The methane is produced by fermentation of organic matter of empirical formula (CH_2O). What mass of organic matter was used in producing the methane in $1 \, dm^3$ of water?

7. (*a*) A waste water contains $300 \, mg \, dm^{-3}$ of biodegradable organic matter, (CH_2O). In passing through a $200\,000 \, dm^3$ a day treatment plant, 40% of the waste is oxidised to $CO_2 + H_2O$. Oxygen is transferred from the air flow into the water with 20% efficiency. Take the percentage of oxygen in air as 20% by volume. Calculate the volume of air at rtp. required for the oxidation.

(*b*) If all the organic matter (CH_2O) were converted into methane, what volume of methane would be produced daily?

5

WATER TREATMENT

5.1 DRINKING WATER

Our water supply comes from lowland surface water, upland catchment water and groundwater.

Each of us uses on average 180 litres of water a day for drinking, washing, cooking and flushing the toilet. All of this water is potable (of drinking water quality). The water which we drink comes from three sources. These are **lowland surface water**, usually from rivers, **upland catchment water** – surface water – which runs off moorlands and mountains, and **groundwater** from underground streams – aquifers – and underground springs. Lowland surface water, e.g. that in rivers, is the dirtiest of the three and needs most treatment before it is safe to drink. The contaminants are agricultural run-off, animal waste, domestic sewage and industrial effluents. Upland catchment water is cleaner because it runs off moorlands and mountains, but it often contains impurities such as acids from peaty soil and agricultural chemicals. Ground-

The different sources need different treatments.

water is often clean enough to drink because it has passed through rocks and soil which filter the water.

The quality of drinking water is governed by the European Directive which lists 66 factors related to quality. In the UK the Department of the Environment has published guidelines on how to maintain hygienic conditions.

5.1.1 TREATMENT OF LOWLAND SURFACE WATER

For lowland surface water, see Figure 5.1A.

Lowland surface water is pumped into storage reservoirs. It is left for 2 to 18 months. Many harmful micro-organisms die off, and some solid particles of sand and silt sink to the bottom. When required it is fed into the water treatment plant [see Figure 5.1A].

Upland catchment water requires less chlorination, but may need a reduction in acidity.

Upland catchment water requires less chlorination than lowland catchment water. It is often acidic because it has passed through peaty earth. The acidity must be neutralised, e.g. by adding lime, to prevent the water corroding iron pipes and lead pipes [see § 6.9.2].

Groundwater needs a modest amount of chlorination.

Groundwater needs chlorination to prevent growth of bacteria, but less chlorine is used. Research is going on into the use of ultraviolet light instead of chlorine to kill harmful bacteria.

Chlorination of humus in water forms contaminants.

In 1970 trihalomethanes, e.g. $CHCl_3$, $CHClBr_2$, were found in drinking water. These compounds are suspected of being carcinogenic. It is believed that trihalomethanes are formed when the water supply is chlorinated and organic compounds in humus [see § 7.7] react with chlorine. Contamination can be reduced by removing as much humus as possible before chlorination.

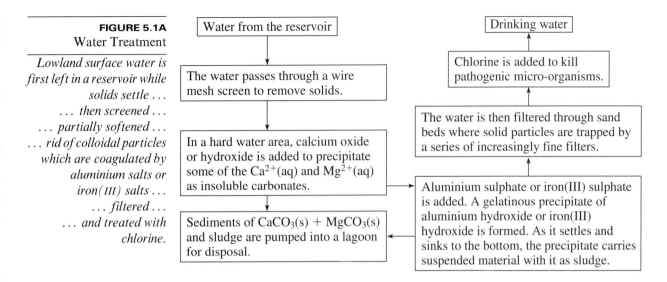

FIGURE 5.1A
Water Treatment

*Lowland surface water is
first left in a reservoir while
solids settle ...
... then screened ...
... partially softened ...
... rid of colloidal particles
which are coagulated by
aluminium salts or
iron(III) salts ...
... filtered ...
... and treated with
chlorine.*

FIGURE 5.1B
Water Treatment Tanks

5.2 FLUORIDATION OF WATER

Tooth enamel consists of calcium hydroxide phosphate, $Ca_5(PO_4)_3OH$, also called
calcium hydroxyapatite. In this ionic structure, an equilibrium exists:

$$Ca_5(PO_4)_3OH(s) + aq \underset{\text{Remineralisation}}{\overset{\text{Demineralisation}}{\rightleftharpoons}} 5Ca^{2+}(aq) + 3PO_4^{3-}(aq) + OH^-(aq)$$

*Calcium hydroxide
phosphate (calcium
hydroxyapatite) in tooth
enamel is attacked by
acids.*

In the environment of the mouth, calcium hydroxyapatite can be dissolved and
reformed. The equilibrium lies to the left. If acids are present in the mouth, however,
they react with hydroxide ions and favour demineralisation. Normally the pH of the
mouth is 6.8. Within the plaque (a gelatinous mass of micro-organisms) which coats
teeth, the pH may be much lower. Sugar is the chief culprit because bacteria in the
plaque convert sugars into acids.

Fluoride ion can replace some of the hydroxide ion in hydroxyapatite. The fluoridated compound is less susceptible to attack by acids.

A relationship has been found between the extent of dental caries (decay) and the concentration of fluoride ion in the drinking water. It has been observed that in cities where the concentration of natural fluoride in the water is high the incidence of tooth decay is lower than average. Fluoride inhibits certain enzymes, such as those that catalyse the fermentation of sugars to lactic acid. Fluoride ion also substitutes for some of the hydroxide ion in calcium hydroxyapatite to form $Ca_5(PO_4)_3(OH)_{1-x}F_x$. The fluoridated hydroxyapatite is less easily attacked by acidic solutions than is hydroxyapatite.

Many people have objected to fluoridation as a kind of compulsory mass medication. Very high levels of fluoride cause damage to teeth. Long term exposure to high levels of fluoride can cause damage to bone, kidney and thyroid. A study by the Royal College of Physicians has concluded, however, that there is no risk to the individual or the environment from levels of fluoride up to 1 ppm. Many water authorities now add sodium fluoride to bring the level of fluoride in drinking water up to 1 ppm. The alternative of taking fluoride tablets is less effective because a large dose of fluoride is rapidly excreted.

CHECKPOINT 5.2

1. For what reasons are the following used to treat water?

(*a*) calcium oxide

(*b*) aluminium sulphate

(*c*) chlorine

(*d*) calcium fluoride

2. How does the fluoridation of drinking water benefit

(*a*) the individual

(*b*) the Government?

3. Give two reasons why fluoridation of drinking water is more effective than issuing fluoride tablets.

4. Why is organic matter removed from water before chlorination?

5.3 PRIMARY TREATMENT OF SEWAGE

Sewage must be treated before being discharged into rivers.

After use, the water passes through sewers and is eventually returned to rivers, from which it is later taken for re-use, or discharged into the sea. Rainwater is usually collected separately and carried to a discharge point, bypassing treatment. The quality of our water supply depends on how well water companies remove pollutants from sewage before discharging it into a river. The treatment of sewage does not vary much from city to city. The treatment of industrial wastes depends on which pollutants are present because different industries give rise to quite different waste products. In the UK the National Rivers Authority monitors discharges into rivers.

Some sewage is discharged untreated into the sea ...
... to be digested by micro-organisms.

When the population is large this method is unsatisfactory.

Many countries, including the UK, discharge part of their sewage untreated into the sea. Eventually micro-organisms is the sea digest it into safe products. The process takes time, and in practice raw sewage floats into coastal waters used by swimmers and pollutes bathing beaches.

Raw sewage can be used as a fertiliser ...
... but it may contain disease-causing organisms and heavy metals.

One method of disposing of raw sewage is to spread it on fields as a fertiliser. This method has been practised for a long time in China. It has the drawback that disease-causing organisms can survive in the soil to be taken up by crops and infect people and animals which consume the crops. Sewage sludge is a safer fertiliser unless it contains other contaminants, e.g. heavy metals [see § 5.6].

The wastewater that flows into the sewers contains floating matter, matter in suspension, colloidal matter and dissolved matter. Other characteristics are pH, colour, odour and the presence of pathogenic micro-organisms. The degree of treatment which wastewater receives depends on its composition and on the standard it must meet before discharge. Primary treatment is illustrated in Figure 5.3A.

Primary sewage treatment involves screening ...

FIGURE 5.3A
Primary Treatment of Sewage.

... settling of sand etc. in a grit chamber ...
... sedimentation of sludge etc. in a sedimentation tank ...

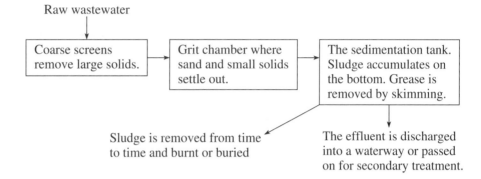

Raw wastewater

Coarse screens remove large solids. → Grit chamber where sand and small solids settle out. → The sedimentation tank. Sludge accumulates on the bottom. Grease is removed by skimming.

Sludge is removed from time to time and burnt or buried

The effluent is discharged into a waterway or passed on for secondary treatment.

FIGURE 5.3B
A Sewage Works

... removal of the sludge ...
... and discharge of the water into a waterway.

Industrial wastes must receive appropriate treatment.

5.4 SECONDARY TREATMENT OF SEWAGE

Secondary wastewater treatment is designed to remove biological oxygen demand, BOD [see §4.15.3]. The principle is to use natural biological oxidation by aerobic micro-organisms (which must be supplied with air) to digest organic material to harmless products: carbon dioxide, water and sludge. Micro-organisms use the oxidation of organic matter to provide the energy which they need to live, and they also use some of the organic matter to synthesise biomass.

Secondary treatment removes biochemical oxygen demand ...

One method of secondary treatment is the **trickling filter** [Figure 5.4A] in which wastewater is sprayed over beds of pebbles covered with a slime composed of micro-organisms. Another type of treatment is the **rotating biological reactors**, plastic discs which bring air into the liquid as they rotate and are coated with a thin layer of micro-organisms. The most effective secondary treatment is the **activated sludge process**

... by supplying air and allowing aerobic micro-organisms to digest organic matter ...

FIGURE 5.4A
Final Sedimentation Tank
at a Sewage Works

... *in a trickling filter* ...
... *or a rotating biological
reactor* ...

FIGURE 5.4B
The Activated Sludge
Process

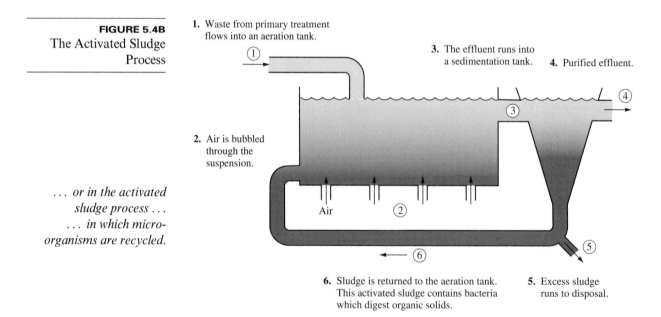

1. Waste from primary treatment
flows into an aeration tank.

3. The effluent runs into
a sedimentation tank.   **4.** Purified effluent.

2. Air is bubbled
through the
suspension.

... *or in the activated
sludge process* ...
... *in which micro-
organisms are recycled.*

Air

6. Sludge is returned to the aeration tank.
This activated sludge contains bacteria
which digest organic solids.

5. Excess sludge
runs to disposal.

[Figure 5.4B]. Micro-organisms that digest organic waste are injected into wastewater
in an aeration tank, which supplies them with plenty of oxygen. The sludge which
settles out contains active micro-organisms and is recycled. Ninety per cent of the
suspended solids and BOD are removed. A means of disposal of the sludge must be
found [see § 5.6].

*Sometimes an anaerobic
secondary treatment is
used to produce methane,
which can be used as a fuel.*

An anaerobic secondary treatment is less widely used. With the lack of oxygen, it
yields methane, carbon dioxide, hydrogen sulphide and inert solids. The methane
produced is used as a fuel in the sewage treatment plant.

*The water is usually
treated with chlorine
before discharge or with
ozone, hydrogen peroxide,
oxygen or UV light.*

Before discharge, the wastewater is disinfected to destroy pathogenic micro-organisms.
Chlorine is commonly used. However, there is concern over the production of
carcinogenic trihalomethanes [see § 5.1] and alternative oxidants have been tried,
including ozone, hydrogen peroxide and oxygen. Ultraviolet light was used in
Cleethorpes, UK, in 1994 to treat sewage on a large scale before discharging it into the
sea. The method is quite common in Australia and the USA.

1. What are the two processes that take place in primary sewage treatment?

2. What is the fundamental principle in secondary sewage treatment?

3. How is methane produced from sewage?

5.5 TERTIARY TREATMENT OF SEWAGE

Tertiary treatment is not widely used.

Some organic compounds and nitrates and phosphates can be eliminated only by **tertiary treatment** or **advanced water treatment**. Tertiary treatment is costly and is not widely used. The methods are described under the treatment of industrial wastewater in § 5.7.

5.6 SLUDGE DISPOSAL

The treatment of sewage sludge involves digestion by anaerobic bacteria and drying, followed by final disposal . . .

Sludges are liquids with a solid content of up to 5.0%, like a thin mud in consistency. A city of 4 million people generates 3 million m^3 a day of sewage, containing 400 tonnes of solid material. Sludge contains many undesirable components. It is treated along these lines:

1. Sewage sludge is digested by anaerobic bacteria, with the formation of stable humus. Harmful micro-organisms are destroyed in the process. In some cases the methane produced can be used as a fuel in the sewage treatment plant.

. . . or incineration – provided air pollution is avoided . . .

2. Sludge is concentrated in sludge drying beds consisting of layers of gravel and sand. More water can be removed by vacuum filtration or centrifugation or by heating.

3. The final disposal may be (*a*) spreading on the land, (*b*) incineration, (*c*) dumping in a landfill or (*d*) dumping at sea.

. . . by spreading on the land as fertiliser – unless it contains heavy metals . . .

(*a*) Sludge can be heated to kill bacteria and then packaged as a fertiliser. It contains about 5% N, 3% P, 0.5% K and humus which improves the physical properties of soil [see § 7.7]. Often the sludge contains high levels of metal ions such as Fe^{3+}, Pb^{2+}, Zn^{2+}, Cu^{2+}, Ni^{2+} and Cd^{2+}, which are toxic to plants. Then the sludge cannot be used as a fertiliser and its nutrient value is lost. If measures are taken to reduce heavy metals in the sewage prior to sludge formation, the sludge can be used more extensively.

(*b*) For incineration the sludge must contain 25% solids if any usable heat is to be generated. Air pollution controls must be in place to avoid atmospheric pollution.

. . . or dumping in a landfill . . .

(*c*) The major part of sludge is dumped in landfills [see § 8.7].

. . . or dumping at sea – which is to be phased out.

(*d*) Dumping at sea is unsatisfactory because there is a tendency for sludge to drift back towards the shoreline. In 1987 the UK dumped about 9000 tonnes of sludge into the North Sea and the Irish Sea. Also dumped in coastal waters were 7000 tonnes of zinc, 3000 tonnes of lead, 1500 tonnes of chromium and 9 tonnes of arsenic. European Community regulations will phase out dumping at sea by 1998.

5.7 INDUSTRIAL WASTEWATER

Industrial wastewater poses somewhat different problems from domestic sewage. Industrial firms are responsible for removing pollutants from their effluent before it is discharged into waterways. In the UK wastewater must meet the standards of the National Rivers Authority.

5.7.1 HEAVY METALS

Industrial wastewaters contain different kinds of pollutants and are treated in different ways, which include …

Copper, cadmium, mercury and lead must be reduced to a low non-toxic level before discharge of wastewater.

PRECIPITATION

Precipitation of metal ions from solution as insoluble hydroxides or basic salts is achieved by adding calcium hydroxide or sodium carbonate.

$$Cr^{3+}(aq) + 3OH^-(aq) \longrightarrow Cr(OH)_3(s)$$

Many metal hydroxides dissolve again at pH > 10 due to the formation of soluble complexes, e.g.

$$Zn(OH)_2(s) + 2OH^-(aq) \longrightarrow Zn(OH)_4{}^{2-}(aq)$$

… precipitation of metal ions, e.g. Ca^{2+}, Mg^{2+}, Fe^{2+}, Mn^{2+} …

Some heavy metals, e.g. mercury, cadmium and lead have insoluble sulphides and can be precipitated by hydrogen sulphide or other soluble sulphides.

ION EXCHANGE

Ion exchange is a means of removing cations or anions from solution on to a solid resin [see *ALC*, §8.7.6]. Heavy metals can be removed from hazardous wastewater.

$$2H^+(resin) + Cd^{2+}(aq) \longrightarrow Cd^{2+}(resin) + 2H^+(aq)$$

Ion exchange is employed in the metal plating industry to purify rinse water and spent plating bath solutions. Cation exchangers remove cations, e.g. Cu^{2+}, Cd^{2+}, and anion exchangers remove anionic complex ions such as $Ni(CN)_4{}^{2-}$ and chromium species such as $CrO_4{}^{2-}$.

ELECTROLYSIS

… removal of heavy metal ions by ion exchange or by electrolysis …

Electrolysis can be used to remove metal contaminants from solution. Both cadmium and nickel can be removed from the wastewater that is produced in the manufacture of nickel–cadmium batteries. Cyanide is often present in the wastewater. If chloride ion is added to the electrolyte, the chlorine generated at the anode oxidises cyanide ion to nitrogen and carbon dioxide.

5.7.2 ORGANIC COMPOUNDS

… removal of organic compounds by activated carbon …

Organic compounds are removed by allowing the water to flow down a bed of activated carbon. Periodically the carbon is heated to a high temperature to oxidise the adsorbed organic componds to carbon dioxide, water, etc. and to regenerate the carbon surface. Dissolved organic compounds which are not adsorbed may be oxidised by ozone, hydrogen peroxide, oxygen or chlorine.

5.7.3 PHOSPHATES

... removal of phosphates by precipitation ...

Phosphates are algal nutrients [see §6.4]. To inhibit the growth of algae, a phosphate level below 0.5 mg dm^{-3} is the aim. Municipal waste may contain 25 mg dm^{-3} of phosphate. The activated sludge treatment removes the organic compounds that contain about 20% of the phosphate from sewage, but the detergent phosphates remain. They can be precipitated by the addition of calcium hydroxide. Another method is absorption on activated alumina.

5.7.4 NITROGEN COMPOUNDS

... removal of ammonium compounds and nitrates.

Nitrogen compounds are algal nutrients. Ammonium ions can be removed by making the wastewater alkaline and passing air through to remove ammonia or by ion exchange. Nitrate ions can be reduced to nitrogen by anaerobic denitrifying bacteria.

CHECKPOINT 5.7

1. (*a*) Name four plant nutrients which sewage sludge contains.

(*b*) When is it not advisable to spread sewage sludge on fields?

2. (*a*) What happens to sewage sludge when it is dumped at sea?

(*b*) Why is this activity to be stopped?

3. How can an industrial firm remove each of the following pollutants from its effluent?

(*a*) cyanide ion

(*b*) organic matter

(*c*) lead(II) ions

(*d*) cadmium ions

4. (*a*) Why is it important to reduce the concentrations of phosphate ions and nitrate ions in wastewater to low levels?

(*b*) State one method for removing phosphate ion.

(*c*) Why is it difficult to remove nitrates?

5.8 ANALYSIS

Water is analysed by titration to find the concentration of acids or the concentration of bases.

Water samples are analysed for their colour, odour, temperature, conductance, turbidity, solid residue after evaporation and other qualities. Samples must be collected and analysis performed as soon as possible. Within a few minutes of collection, the pH may change, dissolved gases may be lost and other gases absorbed from the atmosphere. An analysis of temperature, pH and dissolved gases should be performed in the field. A representative sample of a body of water must be a composite of many samples taken from a number of different locations over a period of time.

5.8.1 pH

In the field a pH probe is used for the purpose.

Acidity and alkalinity can be measured by **titration** [see *ALC*, §12.7.9]. For analysis in the field a **pH probe** is used. This is a glass electrode coupled with a standard electrode in a robust container [see *ALC*, §13.1.3]. Immersed in the water, it gives a direct reading of the pH.

5.8.2 BIOCHEMICAL OXYGEN DEMAND

BOD is measured.

Biochemical oxygen demand is determined as described in §4.15.3. Water is considered pure if the BOD is 1 ppm or less, fairly pure with a BOD of 3 ppm and suspect when the BOD is as high as 5 ppm. The discharge of wastewater with a BOD of 20 into a stream is considered undesirable. Effluent such as the run-off from barnyards, feedlots and food-processing plants can have a BOD as high as 100–10 000 ppm. Such effluents must be diluted many times with huge quantities of water.

5.8.3 GAS CHROMATOGRAPHY

A variety of solutes can be detected and measured by gas chromatography . . .

A mixture of volatile components is carried by a carrier gas through a column containing an adsorbent liquid coated on a solid material [see *ALC*, § 8.7.5]. Each component is partitioned between the carrier gas and the liquid. The components emerge from the end of the column at different times. A detector records the time at which each component emerges and the quantity . The technique can be used to detect a variety of solutes, and quantitative analyses can also be obtained.

5.8.4 HIGH-PERFORMANCE LIQUID CHROMATOGRAPHY

. . . high performance liquid chromatography . . .

The carrier gas is replaced by a liquid which flows through the column. Very high pressures are required to give a reasonable flow rate. The advantage is that the material to be analysed need not be vaporised – a step which may result in decomposition.

5.8.5 MASS SPECTROMETRY

. . . and mass spectrometry.

A mass spectrum is characteristic of a compound and can be used to identify it [see *ALC*, § 1.8]. Computer data banks of mass spectra have been established and can be interfaced with mass spectrometers. The components of a sample can be separated by gas chromatography and the gas flowing out of the gas chromatography column can be sampled continuously by a mass spectrometer. This is a very useful way of analysing organic pollutants. It can be used to identify impurities at less than 1 ppm.

5.8.6 ATOMIC ABSORPTION SPECTROMETRY

Metals can be detected and analysed by atomic absorption spectrometry . . .

Atomic absorption spectrometry is the popular method of analysis for most metals in environmental samples. It is used to detect metals at concentrations down to $1\,mg\,dm^{-3}$ water. The technique is based on the absorption of visible and ultraviolet light. Each metallic element in the sample absorbs a fraction of the radiation that falls on it at a certain wavelength characteristic of the metal. The wavelengths of radiation absorbed identify the metals present in the sample. The fraction of radiation absorbed at a characteristic wavelength indicates the concentration of a particular metal in the sample.

5.8.7 ATOMIC EMISSION SPECTROMETRY

. . . and atomic emission spectrometry.

When metals are heated to a high temperature they emit light. An analysis of the spectrum shows which metals are present. A new technique which enables low concentrations to be analysed depends on raising the temperature in a plasma at 10 000 K in order to increase the emission of light. As many as 30 elements can be analysed simultaneously.

======================= **CHECKPOINT 5.8** =======================

1. An industrial firm wants to check the level of organic matter in its wastewater prior to discharge. How can it do this?

2. A firm wants to find out whether it has been successful in reducing the level of cadmium ions in its effluent to a low level. How can it do this?

3. How can a firm check the concentration of alkali in its effluent?

4. What is BOD? Why is it important to know the BOD of wastewater ? How can it be measured?

======================= **QUESTIONS ON CHAPTER 5** =======================

1. (*a*) Name an ion that is added to water to remove finely divided organic matter.

(*b*) Explain how the addition of this ion leads to the removal of organic matter.

2. Municipal drinking water is obtained from two sources: one a flowing, well-aerated stream with a heavy load of particulate matter and the other an anaerobic groundwater. Describe possible differences in treatment for these two sources of water.

3. (*a*) Why is sludge from the activated sludge process returned to the water treatment compartment?

(*b*) What are the two processes by which the activated sludge process removes soluble carbon compounds from sewage?

4. How does reverse osmosis differ from filtration?

5. Suggest how industries which produce effluent containing the following pollutants can make their wastewater safe to discharge.

(*a*) CN^-, (*b*) Hg^{2+}, (*c*) Pb^{2+}, (*d*) Zn^{2+}, (*e*) PO_4^{3-}, (*f*) Cd^{2+}, (*g*) Ni^{2+}, (*h*) $Ni(CN)_4^{2-}$

6

WATER POLLUTION

6.1 PURE WATER

There is a distinction between 'pure' water, meaning water that is 'fit to drink' and pure water that is a single substance.

Water is an excellent solvent. It is always found in Nature as a solution, never as a pure substance. Natural waters contain micro-organisms as well as solutes. When we describe water as 'pure' we mean that the solutes and micro-organisms in it do not impair health: the water is 'fit to drink'. **Pure water**, distilled water, has a rather 'flat' taste, and is not very pleasant to drink.

A water pollutant is a substance that prevents the use of water for a specified purpose.

Water that is not of drinking quality for human beings may be all right for animals. Water that will sustain fish may be unacceptable for swimming, waterskiing, etc. Water that is too contaminated to sustain aquatic life may be used for cooling in a power plant. Water that is too salty for drinking may be used to irrigate crops. The concept of **polluted water** depends on the use we have in mind for it. We might define a water **pollutant** as any substance that prevents the use of water for a specified purpose.

The signs of polluted water are obvious. Poor drinking water tastes and smells bad. Ponds and slow-running streams are covered with algae. Masses of weeds clog waterways. Nasty odours arise from lakes, rivers and ocean shores. Beaches are stained with oil and littered with rubbish. The flavour of fish is contaminated. Dead fish and decaying vegetation may clog surface waters. Many different substances are classified as pollutants. They are:

1. Pathogens, disease-causing bacteria and viruses

2. Dissolved organic compounds and inorganic compounds

3. Wastes that have a biochemical oxygen demand

4. Nutrients that cause excessive growth of aquatic plants

5. Oily scums and suspended solids and colloids

Many types of substances which are classified as pollutants are listed.

6. Radioactive substances

7. Thermal pollution

6.2 WATERBORNE DISEASES

Diseases are spread by pathogens in impure drinking water.

Diseases spread by impure drinking water include typhoid fever, paratyphoid fever, dysentery, cholera, poliomyelitis and infectious hepatitis. The faeces discharged by a person suffering from one of these diseases may contain large numbers of pathogens – the organisms that cause the diseases. If pathogens get into drinking water used by other

Tests for the presence of pathogens require expert analysis.

people who have not been immunised, an epidemic of the disease will break out. Pathogens do not live for long in water. They may not be detected even if samples are taken at regular intervals. Direct tests for the presence of pathogens require expert analysis. There is a simple method of evaluating them indirectly. All faeces contain a large number of coliform bacteria that live in the large intestine without harming the host. These bacteria are easily detected in a **coliform count**. If they are present, it means that the water is polluted with faeces and might contain pathogens. If they are absent, it means that the water does not contain faecal matter, and pathogens are probably absent.

There is an indirect method of assessing pathogens: the coliform count.

If coliform bacteria are present in a sample of water, the water is polluted with faeces and may contain pathogens. If coliform bacteria are absent, the water does not contain faeces and pathogens are probably absent.

A coliform count is done by making dilutions of the sample and incubating at 35 °C for 48 hours in a lactose nutrient broth. If the medium becomes turbid, this indicates the presence of bacteria. The medium is spread on an agar gel plate and after a further 24 hours the number of bacterial colonies on the plate is counted under a microscope. Even a single living coliform will grow and divide in the culture medium, eventually producing a colony. The accepted standards for coliform counts per 100 cm³ of water are: drinking water < 1, unpolluted lakes 10–100, mild sewage pollution 1000–5000, definite sewage pollution 5000–10 000, dangerous heavy sewage pollution 10 000– 100 000 and sewage > 100 000.

Waterborne diseases have been sharply reduced in countries where water supplies are properly treated. Chlorination of water supplies kills all micro-organisms [see § 5.4]. If water tastes or smells unpleasant, this is due to organic matter having combined with chlorine during the treatment.

FIGURE 6.2A
Their Water Supplies

Waterborne diseases are sharply reduced when water supplies are chlorinated.

Half of the people in Third World countries have no easy access to clean water. Women and children often spend much of their day collecting water from wells and rivers. Often rivers are used for sewage disposal as well as for drinking. Four-fifths of the diseases in the Third World are linked to dirty water and lack of sanitation. Epidemics of cholera, typhoid, and other waterborne diseases still occur in some parts of the world. Diarrhoea kills 5 million children every year. The technology of eliminating waterborne diseases is simple and not costly. The United Nations named 1980–90 as the Water Decade. The aim was to achieve clean water and sanitation for all by 1990. Progress was made by investment in pumps and disinfection kits, but the goal is still far from being achieved.

6.3 BIOCHEMICAL OXYGEN DEMAND

The dissolved oxygen, DO, in water is used for respiration of aquatic plants and animals.

The **dissolved oxygen, DO**, in water is used for respiration by aquatic plants and animals. Any added material that uses up dissolved oxygen therefore interferes with the growth of aquatic organisms. Fish require the highest concentration of oxygen, followed by invertebrates – protozoa, worms, clams, shrimps, insects, etc. – and bacteria require the least oxygen. If the dissolved oxygen falls below 5 ppm, fish are the first to suffer and tend to die out. The populations of invertebrates and bacteria then rise to abnormal levels. This imbalance between species is a sign of pollution.

Substances which use up dissolved oxygen and add to the biochemical oxygen demand, BOD, are pollutants.

Such substances come from human and animal wastes, food canneries, meat packaging plants, etc.

In §4.15.2 it was calculated that 8 mg of decaying organic material uses up all the dissolved oxygen in 1 dm³ of water. Such organic pollutants are present in human and animal wastes and in effluents from food canneries, meat packaging plants, slaughterhouses, paper mills, tanneries and chemical factories. If the level of oxygen becomes too low to support aquatic life, algae and other plants die and decay, thus adding to the demand for oxygen. The amount of dissolved oxygen used up during oxidation by bacteria of the organic matter in a sample of water is called the **biochemical oxygen demand, BOD** [§4.15.3 and 5.8.2]. Water is rated as pure if the BOD is 1 ppm or less, fairly pure with a BOD of 3 ppm and suspect when the BOD reaches 5 ppm.

CHECKPOINT 6.3

1. What is the difference between pure water and unpolluted water?

2. Briefly describe a coliform count. Does the presence of coliform bacteria prove that water is contaminated? Does the absence of coliform bacteria prove that water is uncontaminated?

3. What is the solution to the problem of unsafe drinking water?

4. (*a*) Define 'biochemical oxygen demand' and 'chemical oxygen demand'.

(*b*) Describe how biochemical oxygen demand is measured.

6.4 EUTROPHICATION

Lake water is poor in plant nutrients and supports little plant life. With no decaying vegetation to use up oxygen in lake water, fish thrive.

In order to grow, green plants require about 20 elements. The elements carbon, hydrogen, oxygen, nitrogen and phosphorus are needed in substantial amounts. In normal conditions, water always provides enough carbon, hydrogen and oxygen for plant growth. The rate of growth is limited by the supplies of nitrogen and phosphorus. Lake water, being poor in nitrogen compounds and phosphorus compounds, supports little plant life. There is no decaying vegetation to use up dissolved oxygen, the lake water is saturated with oxygen, and fish can thrive.

Lake water may be enriched with nutrients – a process called eutrophication – which encourages plant growth and leads slowly to ageing.

Under natural conditions, streams flow into the lake bringing compounds of nitrogen, phosphorus and sulphur which are formed from the decay of plant and animal refuse on land. Calcium compounds are leached (dissolved by percolation of water through the soil) from the soil and enter the lake water. The lake water is gradually enriched with nutrients, a process called **eutrophication**. Water plants grow more vigorously, animal populations increase, then as plants and animals die deposits of organic matter on the bottom of the lake build up. As the water becomes shallower it becomes warmer and plants and animals grow faster. Plants take root in the shallows and the lake becomes edged with marshland. In time the marsh is overrun by land plants and the lake is completely filled in. This natural process of **ageing** takes thousands of years.

Ageing may be accelerated by human activities . . .

Ageing is accelerated when plant nutrients are fed into a lake by human activities. When a lake contains concentrations of nitrates and phosphates higher than normal, algae flourish and produce a bloom [see Figure 6.4A], a green scum which is accompanied by an unpleasant odour and taste in the water. When the bloom of algae dies it is decomposed by **aerobic bacteria**. When the oxygen content is insufficient to support aerobic bacteria, **anaerobic bacteria** take over. They convert the dead matter into unpleasant-smelling decay products and debris which fall to the bottom. Gradually a layer of dead plant material builds up on the bottom of the lake. The lowering of the oxygen concentration leads to the death of fish and the replacement of fish by species which can survive in water with a low oxygen concentration, such as eels and sludge worms. The loss of fish is a disaster for fishermen. The mats of algae ruin the lake for recreational use by swimmers, sail-boarders, dinghy-sailors, rowing boats and yachts.

. . . through the leaching of fertilisers from the surrounding land and through wastewater containing detergent phosphates . . .

The sources of the nitrates and phosphates are sewage and fertilisers. Intensively cultivated land receives generous applications of fertilisers containing nitrates and phosphates. Plants can absorb only a limited quantity of nitrate through their roots. The rest is leached out of the soil by rain. Nitrates are very soluble, phosphates are sparingly soluble and are leached from the soil more slowly than nitrates. Another source of phosphates is household detergents.

FIGURE 6.4A
Algal Bloom

. . . causing eutrophication, algal bloom, the build-up of dead matter on the bottom of the lake . . .
. . . and the death of fish.

6.4.1   NITRATES

Excessive fertiliser can also find its way into **groundwater**. This is the water held underground in porous layers of rock. One-third of UK drinking water comes from groundwater. The World Health Organisation recommends that the level of nitrogen in the form of nitrates should not exceed 50 ppm. The average level in the UK is far short of this value (about 11 ppm), but it is increasing and in some parts of the country where the level is highest there is concern.

The level of nitrates in groundwater is rising . . .

The worry over nitrates is twofold. One is that nitrates are converted into nitrites, and nitrites oxidise the iron in haemoglobin from iron(II) ions to iron(III) ions. In the oxidised form haemoglobin can no longer transport oxygen round the body. Babies are more at risk than adults because the lower acidity in babies' stomachs encourages the conversion of nitrates into nitrites. The extreme case of nitrite poisoning is the very rare 'blue baby' syndrome, in which the baby turns blue from lack of oxygen. The second worry is that nitrosoamines have been found in a number of foods. They are carcinogenic, and some chemists think that they may have originated from nitrites formed from nitrates.

. . . giving rise to concern over nitrite poisoning and carcinogenic nitrosoamines.

Some progress has been made on reducing the run-off of nitrates from agricultural land. Agricultural chemists are advising farmers how they can make the most efficient use of fertilisers and avoid using an excess which the plants cannot take in. The correct quantity applied in the right season will be taken up by the growing plants. This saves the farmers money as well as cutting down on pollution. For removal of nitrates from water see § 5.7.4.

It is important to apply nitrate fertilisers economically.

6.4.2 PHOSPHATES

Phosphates enter water from fertilisers and also from detergents.

Phosphates enter water from fertilisers, but they are not leached out of the soil as rapidly as nitrates because they are less soluble. Soapless detergents use phosphates as brighteners, and detergents enter the sewers and are discharged into rivers. This source of pollution could be avoided if people would settle for a detergent which left their shirts just a shade less white. Then phosphates could be omitted from detergents and stop polluting lakes, rivers and groundwater.

Phosphates are added as 'builders' to improve the cleaning power of synthetic detergents.

Why are phosphates used in detergents? Most soaps do not work well in hard water because the calcium ions and magnesium ions in the water convert the soap into an insoluble scum of calcium and magnesium salts. One way of maintaining the soap in the form of soluble anions is to add **builders**, e.g. sodium phosphate. These builders raise the pH however, possibly as high as pH 12, and make the water rather caustic to the skin and also accelerate the rate of wear of articles being washed.

The first synthetic ABS detergents were not biodegradable.

Synthetic detergents have the advantage that their calcium and magnesium salts are soluble. The alkylbenzenesulphonates (ABS) were first used in the 1940s, and their use grew rapidly when it was found that they reacted synergistically with the phosphate builders in the washing powders. (A synergistic effect means that each component works better in the mixture than it does separately.) The synergistic effect meant that cleaning could be achieved with smaller quantities of synthetic detergents. Detergents took over from soaps, and much larger quantities of phosphate builders began to be consumed.

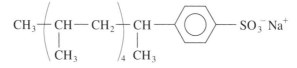

ABS detergent

LAS detergent

Unfortunately the ABS detergents were not biodegradable, so streams became covered with detergent foam, and public objection was strong. In the 1960s, biodegradable linear alkylbenzenesulphonate (LAS) detergents were developed. They still required the addition of large quantities of phosphates or other builders for maximum cleaning power at minimum price.

The LAS detergents are biodegradable but still require the addition of phosphates.

Large quantities of phosphates are released into sewage treatment plants and, after hydrolysis to $H_2PO_4^-$, released into natural waters. The public reaction to eutrophication in the 1970s prompted detergent manufacturers to look for ways of reducing the use of phosphates. The approach was threefold:

- a search for new builders,

- the use of known less effective builders, e.g. carbonates and silicates. These builders give solutions with pH values as high as pH 13, which are very dangerous if accidentally swallowed by children.

The solution to the problem is to find other 'builders' or to remove phosphates by tertiary treatment.

- removal of phosphate at the sewage works by tertiary treatment [§ 5.7].

So far no one approach has solved the problem.

CHECKPOINT 6.4

1. What factor is it that most frequently limits the rate of plant growth?

2. Why is a lake a poor environment for plants and a good environment for fish?

3. Describe the conditions under which a lake becomes more able to support plant growth. Explain why this is not environmentally beneficial.

4. Why are nitrates used in fertilisers? What measures can be taken to prevent nitrates used as fertilisers finding their way into lakes and groundwater?

5. (*a*) Why are phosphates used in fertilisers and in detergents?

(*b*) Suggest what can be done to reduce the build-up of phosphates in waterways.

6.5 DISSOLVED ORGANIC COMPOUNDS

Natural organic compounds have been present in the environment for millions of years, and living organisms have adapted to using some of them and tolerating others. Many synthetic chemicals are harmless to living things, but some interfere with biochemical processes. Some of these synthetic chemicals are purposely introduced into the environment to kill insects, weeds and other unwanted organisms. Others enter the environment by accident. Millions of tonnes of synthetic organic compounds are manufactured, and much escapes into industrial waste either by accident or as a result of dumping. Many of these compounds are not biodegradable; they cannot be broken down by bacteria or other organisms so they accumulate in the environment. Some, like DDT, are concentrated by passage through a food chain [see § 8.4]. Bacteria and protozoa ingest the compound and are then eaten by larger organisms.

Some organic compounds pollute waterways. They are especially dangerous if they are not biodegradable.

These larger organisms are eaten by still larger organisms, and at the end of the food chain fish in the seas and animals on land carry concentrations of DDT much higher than the level in the environment as a whole. Many of the land pollutants mentioned in Chapter 8 are also water pollutants because they are leached out of the soil into waterways. For the removal of organic compounds from water see § 5.7.

6.6 ACIDIC POLLUTANTS

Mines and mine tips are a source of pollutant acids.

A source of acidic inorganic pollutants is the drainage of water from mines and mine tips. Iron pyrite, FeS_2 is a common mineral. It is acted on by bacteria in the presence of air and water to give a solution of sulphuric acid and iron(II) sulphate. This acidic solution makes its way into groundwaters and emerges in springs and streams. Water flowing from underground mines is usually highly acidic. Surface waters running from mine tips are likely to be similarly polluted. A large fraction of this pollution comes from abandoned mines.

Most natural freshwaters are nearly neutral, some are slightly acidic and others are slightly alkaline. The chemical reactions which take place in plants and animals

In lakes and rivers, the acids react with carbonate ions and hydrogencarbonate ions to form carbon dioxide.

proceed within these limits. When acidic pollutants are dumped into natural waters, these processes are upset. Reactions take place with dissolved carbonate ions and hydrogencarbonate ions to produce carbon dioxide.

$$CO_3^{2-}(aq) + H^+(aq) \longrightarrow HCO_3^-(aq)$$

$$HCO_3^-(aq) + H^+(aq) \longrightarrow CO_2(aq) + H_2O(l)$$

Aquatic animals produce carbon dioxide in their metabolism. It is carried in the blood to the respiratory organs, e.g. the gills of fish, which are bathed in water. If the level of dissolved carbon dioxide in the water is high, carbon dioxide in the animal's blood cannot readily diffuse into the water. Carbon dioxide accumulates in the blood, interfering with the transport of oxygen to the tissues, and the animal dies. Exposure to acidic water kills plants as well as animals.

A raised level of carbon dioxide in the water makes it more difficult for aquatic animals to respire.

Extreme pollution by acids has the following effects:

1. The destruction of plants, with the exception of some bacteria and algae

2. The death of aquatic animals – all vertebrates, most invertbrates and many micro-organisms.

Extreme acid pollution kills aquatic plants and animals, corrodes metal structures and damages irrigated crops.

3. Corrosion of structures such as metal pipelines, concrete bridges, iron piers, iron gates in canals and locks, metal ships

4. Damage to irrigated crops

One method of preventing pollution by acidic mine water is to add solid calcium hydroxide to produce a muddy slurry. Sulphuric acid is neutralised.

$$Ca(OH)_2(s) + H_2SO_4(aq) \longrightarrow CaSO_4(s) + 2H_2O(l)$$

Run-off from mines can be treated with calcium hydroxide.

Air is bubbled through the slurry, and hydroxide ions from the calcium hydroxide precipitate iron as iron(III) hydroxide.

$$Fe^{3+}(aq) + 3OH^-(aq) \longrightarrow Fe(OH)_3(s)$$

The slurry is pumped into a lagoon where the solids settle out. The purified water is run into streams. The problem of how to dispose of the sludge in the lagoon remains.

For acid rain see § 3.6. The effects of acid rain are discussed in §3.6.

6.7 CYANIDES

Cyanides are widely used in industry for cleaning metals and in electroplating. They are used in mineral-processing operations, e.g. leaching gold from gold ores by combining to form a soluble complex ion. At one time firms discharged cyanide waste into streams and rivers. This practice is now prohibited, yet sometimes cyanides escape into waterways. In water, cyanide ion sets up an equilibrium:

$$CN^-(aq) + H_2O(l) \rightleftharpoons HCN(aq) + OH^-(aq)$$

The weak acid hydrocyanic acid is formed. It comes out of solution as the extremely poisonous gas hydrogen cyanide. As hydrogen cyanide leaves the system, the equilibrium shifts from left to right.

Cyanides enter waterways in effluents from industries which use them for cleaning metals, electroplating, etc.

The toxic effect of cyanide is due to the formation of a complex with iron(III) ions, preventing the reduction to iron(II) which is part of the chain of reactions by which the body uses oxygen in respiration. The gas is so poisonous that it is used for executions in some states of the USA. For the removal of cyanides from water see §5.7.

6.8 ALUMINIUM

Aluminium ions are present in the water supply because aluminium sulphate is used in water treatment.

Aluminium in drinking water is mainly derived from treatment with aluminium sulphate (alum) [see § 5.1]. Aluminium remaining in the water is removed by treatment, but some of it passes into the water supply. Some parts of the country have higher concentrations than others. A survey was made in 1989 of the incidence of Alzheimer's disease and the concentration of aluminium in drinking water. There is evidence of a correlation. The rate of Alzheimer's disease in districts where the concentration of aluminium in drinking water exceeds $0.11\,\text{mg}\,\text{dm}^{-3}$ is higher than the rate in districts where the aluminium concentration is less than $0.01\,\text{mg}\,\text{dm}^{-3}$. The researchers believe that there is a cause and effect relationship between aluminium concentration and Alzheimer's disease. However, other factors, in addition to aluminium concentration, vary between different regions, and the results of the survey are being interpreted with caution while investigation continues.

There is evidence of a correlation between the aluminium content of drinking water and the incidence of Alzheimer's disease.

In the 1970s it was observed that many patients on long-term kidney dialysis developed Alzheimer's disease. Large volumes of water are used in dialysis, thus exposing patients to higher amounts of aluminium than normal. In addition, the patients were given large doses of antacids containing aluminium oxide. When the connection was made between dementia and aluminium, aluminium was removed from the dialysis fluid and there was a decline in new cases of the disease.

Dialysis patients have in the past been exposed to high levels of aluminium ion and many have developed Alzheimer's disease.

Alzheimer's disease is a form of senile dementia. An early symptom is loss of short-term memory. Sufferers lose their physical and intellectual powers gradually over a period of 10–20 years. The disease can strike the most brilliant intellectuals and can begin in people aged 50, although it usually strikes people in their seventies.

CALAMITY IN CORNWALL

There was an accidental discharge of aluminium sulphate into the water supply in Camelford.

In Camelford in Cornwall in July 1988, a truck driver made a mistake. He arrived at a water treatment plant with a load of 20 tonnes of an 8% solution of aluminium sulphate. There was no-one manning the plant so the driver decided to pump his cargo into a tank. Unfortunately, it was the wrong tank, and the aluminium sulphate went into the water supply for 20 000 people. It was six weeks before the truth was admitted by the South West Water Authority. The public was at first assured that there was no danger, but a hundred residents became ill with nausea and burnt mouths, vomiting and diarrhoea. The acidic solution attacked water pipes, releasing lead, copper and zinc into drinking water, with potentially harmful effects for infants and unborn children. One effect of the copper salts was that people with blonde hair who washed it in the contaminated water saw their hair turn green. There is a more serious side to the matter in that copper in drinking water should not exceed $3\,\text{mg}\,\text{dm}^{-3}$, and the Camelford water had seven times this concentration. Copper poisoning can lead to damage to kidney, brain, liver and eyes. If there is a causal relationship between aluminium in drinking water and Alzheimer's disease, more damage may be revealed in the future.

=================== CHECKPOINT 6.8 ===================

1. (*a*) Name one source of acidic pollution in waterways.

(*b*) What dangerous effect can an acidic medium have on fish?

2. (*a*) Name one source of cyanides in waterways.

(*b*) What can happen if water contains both acidic pollutants and cyanides?

3. (*a*) Name one source of aluminium in drinking water.

(*b*) Explain why there is concern over high levels of aluminium ion in water.

6.9 HEAVY METALS

6.9.1 CADMIUM

Heavy metals are serious water pollutants.

Cadmium comes from the waste from zinc mines, from metal plating and pigments.

Pollutant cadmium in water may come from industrial discharges and the waste from zinc mines. Cadmium and zinc are common both in water and in sediments in harbours surrounded by industrial installations. Cadmium is widely used in metal plating; and in an orange pigment used in paints and enamels. Cadmium is highly toxic, with a recommended upper limit in drinking water of only 10 ppb. It causes high blood pressure, kidney damage and destruction of red blood cells. Some of its effects may arise from the replacement of zinc in some enzymes by cadmium, which is in the same sub-group of the Periodic Table as zinc.

6.9.2 LEAD

Lead compounds are pollutants of air, water and land. Leaded petrol is a source of lead in the atmosphere [see § 3.2]. Particles of soot and lead compounds from vehicle exhausts can fall on land and contaminate crops. From the land, lead compounds can be leached into rivers and groundwater.

Lead comes from lead compounds in the atmosphere, from lead pipes and solders, from lead glazes on pottery and glasses.

Some natural waters contain lead which has dissolved from lead ores, which are often present in limestone. Lead pipes and lead solders were used in plumbing until the 1950s. Since then lead pipes have been replaced by copper pipes because these are easier to make and to bend. More recently cold-water systems have used plastic pipes which are cheaper. In many parts of the UK there are houses which are old enough to have lead plumbing. In hard-water areas this is not dangerous. Lead pipes acquire a lining of insoluble calcium carbonate, magnesium carbonate and lead(II) carbonate. This lining prevents lead from dissolving. In soft-water areas, the pH of the water may be as low as 5, and lead dissolves from lead pipes. The World Health Organisation recommends that the limit for lead in drinking water should be $50 \, \mu g \, dm^{-3}$. Water Companies increase the pH of acidic waters by adding calcium hydroxide. During the 1983 strike of UK water workers, the water was not treated and in houses with lead pipes the level rose in 2 weeks from $30–40 \, \mu g \, dm^{-3}$ to $800–1200 \, \mu g \, dm^{-3}$. Even copper pipes can lead to contamination if the joints are made with lead solder.

The use of lead glazes on pottery and glasses has been a source of lead poisoning in the past. There have been incidents of lead compounds dissolving in acidic drinks such as fruit juices. In 1995 a regular drinker of cider sued the landlord of her 'local' for serving her cider in a glazed ceramic mug. Lead compounds from the glaze dissolved in the acidic liquid and gave her headaches and depression. There is now legislation to restrict the amount of lead that may be released from glazes on dishes that are used for food and drink to 7 ppm. There is evidence that the amount of lead present in the human body has decreased over the last few decades.

6.9.3 MERCURY

Mercury compounds are found in nature in low concentrations in rocks and soils, in still lower concentrations in river water and in minute amounts in air. Living organisms are not affected by these concentrations. However, we find a variety of uses for mercury which add mercury to the environment. The total quantity of mercury released into the environment can be substantial.

1. The use of mercury cathode cells for the electrolysis of brine to give sodium hydroxide and hydrogen [*ALC*, § 18.5.5] should in theory involve no loss of mercury.

In practice the loss of mercury is substantial. The loss can be reduced by running waste water into lagoons where mercury settles out . The alternative diaphragm cell avoids the use of mercury altogether [*ALC*, § 18.8.2].

2. Mercury compounds are used as fungicides on seed grains. These compounds can be absorbed by plants that sprout from the seed and passed on to animals that eat the plants.

3. Paints formulated for use on ships use mercury compounds.

4. Coal contains about 1 ppb of mercury. Although the proportion is small, the quantity of mercury released into the atmosphere from burning fossil fuels is estimated to be over 5000 tonnes per year.

Mercury comes from mercury cathode electrolysis cells, from fungicides, paints, coal and disinfectants.

5. Sludge from sewage treatment plants contains about 1 ppb of mercury. It comes from pharmaceuticals, disinfectants and paints. About 500 kg of mercury passes into the sewers each year from a population of 1 million people. Sewage effluent sometimes contains 10 times the level of mercury found in natural waters.

The toxic nature of mercury was illustrated in Minamata Bay in Japan in 1953–60. There were 43 deaths and 111 cases of mercury poisoning. The illness was traced to seafood caught in Minamata Bay. The fish had been contaminated by mercury waste from a chemical plant that drained into the bay. Mercury poisoning results in slurred speech, numbness, loss of hearing and vision, disorientation and constant trembling.

Minamata is a fishing village on the shore of Minamata Bay in Japan. In 1951 a plastics factory started discharging waste into the bay. By 1953, a thousand people in Minamata had become ill. Some were crippled, some were paralysed, some lost their sight and others their mental abilities. Many died.

The concentration of mercury compounds in the bay water was low (2 ppb), and scientists could not understand how it could poison people. Then they realised that the mercury compounds became part of a food chain [see Figure 6.9A].

FIGURE 6.9A
The Food Chain which Led to Minamata Disease

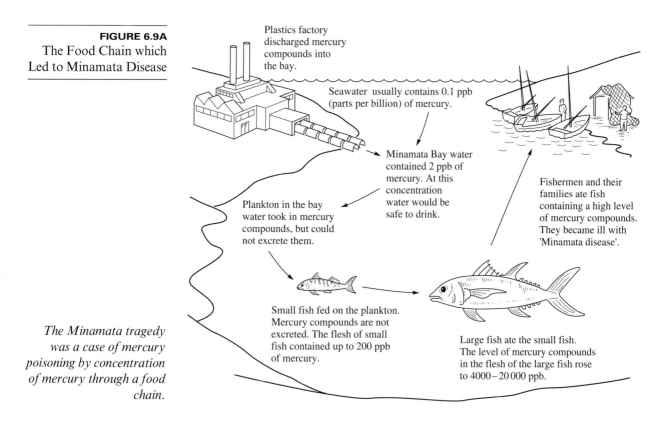

Plastics factory discharged mercury compounds into the bay.

Seawater usually contains 0.1 ppb (parts per billion) of mercury.

Minamata Bay water contained 2 ppb of mercury. At this concentration water would be safe to drink.

Plankton in the bay water took in mercury compounds, but could not excrete them.

Fishermen and their families ate fish containing a high level of mercury compounds. They became ill with 'Minamata disease'.

Small fish fed on the plankton. Mercury compounds are not excreted. The flesh of small fish contained up to 200 ppb of mercury.

Large fish ate the small fish. The level of mercury compounds in the flesh of the large fish rose to 4000–20 000 ppb.

The Minamata tragedy was a case of mercury poisoning by concentration of mercury through a food chain.

Mercury has accumulated in a number of lakes. It is slowly converted into dimethylmercury and methylmercury which are soluble ...

In spite of this terrible experience, other countries have been slow to deal with mercury pollution. In 1970 Canada and the USA found that hundreds of lakes had mercury levels that made fishing dangerous. The mercury and mercury compounds at the bottom of those lakes are still there; they are being slowly converted into toxic methylmercury [see below]. In 1988 the ICI plant on Merseyside discharged more mercury than the Water Authority permits.

... can be ingested by animals ...

Inorganic compounds of mercury tend to concentrate in the tissues of the liver and kidneys. Mercury is rapidly removed from these organs by excretion in urine, but the damage to the organs remains. If mercury vapour is breathed into the lungs, it is quickly picked up by the blood and concentrated in the brain, where it can do serious damage. When regulated doses of mercury compounds are used as medicines, they are quickly excreted by the body, and little mercury accumulates in the tissues.

... and be concentrated in fatty tissues.

The most dangerous mercury compounds are organic compounds. Those which contain methylmercury ions, CH_3Hg^+ and dimethylmercury, $(CH_3)_2Hg$, are the most deadly. When mercury enters a stream or lake it settles into the mud at the bottom. There it is changed by anaerobic micro-organisms into dimethylmercury. This compound soon dissolves and is changed into methylmercury which is readily absorbed by aquatic animals and concentrates in their fatty tissues. By concentration through the food chain, as described above, it can kill people who eat fish containing large amounts of methylmercury.

6.9.4 CHROMIUM

Chromium(VI) compounds are toxic but are soon reduced to harmless chromium(III) compounds.

Chromium(VI) compounds are irritating and corrosive when swallowed or inhaled or absorbed through the skin. They are present in the wastes from electroplating plants and tanneries. Fortunately chromium(VI) is rapidly reduced in aqueous solution to chromium(III), which shows little toxicity.

6.9.5 NICKEL

Nickel compounds come from electroplating factories and cause damage to lungs and brain.

Nickel has carcinogenic properties and also many people suffer an allergic reaction to it. Acute poisoning causes dizziness, vomiting, coughing and shortness of breath. The damage in acute cases is to the lungs and brain. Nickel compounds are present in the effluent of electroplating factories.

6.9.6 COPPER

Copper(II) compounds are toxic.

Copper(II) ions can react with —SH groups in proteins. If the intake is high enough, it can damage kidney, brain, liver and eyes.

6.9.7 ZINC

Zinc enters the water supply from galvanisng plants.

Zinc is relatively non-toxic, but large doses cause vomiting and diarrhoea. Zinc can enter the water supply from the effluent of galvanising plants.

6.10 ARSENIC

Arsenic is present in many rocks and frequently in phosphates so it occurs as an impurity in fertilisers and detergents. Contamination by agricultural pesticides, e.g. lead arsenate and sodium arsenite, has occasionally been a source of arsenic poisoning. The combustion of fossil fuels sends into the atmosphere each year 5000 tonnes of arsenic, which is eventually precipitated on the ground and leached into waterways. The level of arsenic in natural water is about 2 ppb. The recommended upper limit for drinking water is 50 ppb. Arsenic is concentrated in food chains and is converted by bacterial action into soluble, toxic compounds of dimethylarsenic, $(CH_3)_2As^+$.

Arsenic is not a 'heavy metal'. It comes from the combustion of fossil fuels and is concentrated in food chains.

Ingestion of 100 mg of arsenic by an adult is fatal. Chronic poisoning is caused by the ingestion of small quantities of arsenic over a long period. Symptoms of arsenic poisoning are weakness, vomiting and diarrhoea, followed by loss of sensation in feet and hands and, in severe cases, death. The effects of arsenic are caused by the coagulation of proteins due to its reaction with —SH groups and the inhibition of enzymes which catalyse the production of adenosine triphosphate, ATP.

CHECKPOINT 6.10

1. Name a source of cadmium in water.

2. One source of lead in water is lead water pipes. Explain why the danger is reduced in hard-water regions.

3. How does leaded petrol affect the level of lead in water supplies?

4. (*a*) Name one source of mercury in water supplies.

(*b*) Mercury is an insoluble element, so how can it be readily absorbed by animals?

(*c*) Explain why it is difficult to reduce the use of mercury.

(*d*) Mention two symptoms of mercury poisoning.

5. Name industries which gives rise to effluents containing:
(*a*) chromium
(*b*) nickel
(*c*) zinc.

6. (*a*) Name two sources from which arsenic might contaminate waterways.

(*b*) What are the symptoms of arsenic poisoning?

6.11 OIL SLICKS AT SEA

Millions of barrels of crude oil are transported from mines to refineries each year. Modern tankers hold up to 500 000 tonnes of crude oil.

Sometimes there are accidents: tankers may collide or they may scrape a rock.

Millions of barrels of crude oil are pumped from the earth every year. The oil is transported to refineries through pipelines, in tanks on ships, in tanks on trains and in trucks. The refinery products, petrol, diesel oil and others, are transported to customers by rail and road. With all this transporting of oil and fuels, there are sometimes accidents. The most serious accidents happen with oil tankers because they are so large. The modern oil tankers hold up to 500 000 tonnes of crude oil. Large slicks (floating islands of oil) have been created on oceans by the wrecking of tankers and by leakage of oil from oil wells drilled in shallow seas. Sometimes the pollution is deliberate. Tankers have been known to wash out their tanks at sea to save time in port. This is illegal. The total is 4–5 million tonnes of oil added to the ocean each year.

FIGURE 6.11A
Clearing an Oil-polluted
Beach after a Tanker
Accident

*When accidents happen,
tonnes of oil pollute the
ocean to form giant oil
slicks. In addition, illegal
washing out of tanks at sea
causes oil slicks.*

6.11.1   THE *EXXON VALDEZ*

The trans-Alaska pipeline carries oil from wells in the north of Alaska to the port of
Valdez [see Figure 6.11B]. Tankers fill up at Valdez and sail south to deliver the oil to
the rest of the USA. On 24 March, 1989 a supertanker called the *Exxon Valdez* left
the port carrying 215 000 tonnes of oil. Only 40 km out of port, the ship struck a
submerged reef, Bligh Reef, which punctured some of the ship's oil tanks. From the
holed supertanker leaked 36 000 tonnes of oil to form a slick of 1300 square kilometres
in a shallow area called Prince William Sound.

FIGURE 6.11B
Map of Alaska

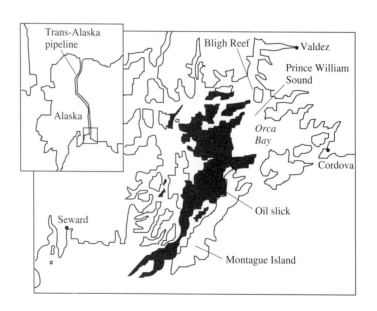

In 1989 the supertanker
Exxon Valdez *struck a reef
and leaked 36 000 tonnes of
oil which polluted an area
of great natural beauty and
great importance to
wildlife.*

Ten hours passed before Exxon Shipping, the company who own the ship, tried to
stop the spill spreading. By that time, high winds and choppy seas made it difficult to
apply any of the methods of dealing with oil slicks. The slick moved out of Prince
William Sound and reached the shoreline. Exxon Shipping and Alyeska, the company

which operates the oil terminal at Valdez, recruited 600 people to clean up the mess. They scrubbed 560 km of oil-soaked coastline, blasting rock faces with hot water under pressure and washing sand and gravel with cold water. They suspended their clean-up operation in September 1989, annoying the state of Alaska which had to spend $21 million on further cleaning.

The owners carried out a clean-up operation, but thousands of sea-mammals and thousands of seabirds died.

The effects on marine life were disastrous. Thousands of sea-mammals – sea lions, seals and sea otters – suffered from the pollution. Sea otters [see Figure 6.11C] that become covered with oil have no chance of recovery because they keep warm by means of a coat of fur which is protected by a layer of long guard hairs. If the guard hairs become soiled, the fur underneath becomes wet, cannot trap air and keep the animal buoyant, and the otter drowns. In spite of a rescue operation, 10 000 sea otters and 16 whales had died by September 1989.

FIGURE 6.11C
A Sea Otter

Every year in May, 15 million migrating birds arrive in Alaska. Waders, 11 million of them, stop on their way home from wintering in the south in the shallow water of the shoreline near Prince William Sound. Seabirds began to arrive while the oil spill clogged the open water where they feed on fish and plankton. By mid-September 1989, 34 500 seabirds had died.

There have been other accidents too, which were serious because of the large size of tankers ...

... and a double hull has been proposed as a safety measure.

The *Exxon Valdez* disaster was by no means the biggest ever. The increase in the size of tankers since 1945 has been spectacular. In 1945 the deadweight of the largest tanker was 16 500 tonnes, whereas the *Seawise Giant* built in 1980 has a deadweight of 565 000 tonnes. The very large crude carriers are difficult to stop and to change direction, and when accidents happen the results are more serious. In 1967 the wreck of the *Torrey Canyon* had discharged 120 000 tonnes of oil off the south-west coast of England. In 1978 the wreck of the *Amoco Cadiz* had sent 220 000 tonnes of oil onto the beaches of Brittany in France. The record is the *Atlantic Express* which lost 270 000 tonnes of oil off Trinidad in the Caribbean in 1979. The *Exxon Valdez* accident promoted legislation to prevent repetition of such disasters. In 1990 the Oil Pollution Act was passed in the USA, specifying that in future tankers that want to operate in US waters must have a double hull, with a 3 m gap between the outer hull and the oil tanks inside. Tanker owners are liable for compensation for damage that oil spills cause. A double hull is not the answer to all accidents. It would not have helped when the *Amoco Cadiz* broke in two, and it will not withstand high-speed collisions. It will help in the more common type of accident where a tanker scrapes over a rock. It would have saved the day when the *Sea Empress* grounded off the UK in 1996.

6.11.2 THE *SEA EMPRESS*

In February 1996, the oil tanker *Sea Empress* ran on to rocks outside Milford Haven. A stretch of Welsh coastline 190 km long was contaminated by oil. Oil slicks hit the marine nature reserves of Skomer and Lundy Islands and six sites of special scientific interest (SSSIs on the map; Figure 6.11D). These include major bird colonies and the country's only coastal national park. Within a month, over 1000 oiled birds had been counted, and 300 dead birds washed ashore. The oil reached Grassholm where 65 000 gannets were returning after the winter to their breeding ground. Guillemots are especially vulnerable because these sea birds dive to avoid danger, rather than fly away, and when they resurface under a film of oil their feathers become matted together.

The oil tanker Sea Empress *grounded on rocks off the coast of Wales in 1996. The leakage of crude oil was estimated as 70 000 tonnes.*

When the *Sea Empress* rammed a sandbank off the coast of Wales on 15 February, 1996, she was carrying 130 000 tonnes of crude oil on her way to Milford Haven refinery. After losing 6000 tonnes of oil she was refloated by tugs. The next day the weather was good, the ship was afloat and held in position by four tugs. The salvage team began planning how to transfer the remaining oil to other vessels. On 17 February, the tanker's pump-room flooded, delaying the pumping out of oil, and the ship grounded again. Storms were forecast. Pilots called for the ship to be towed out to sea where the tides would be less strong and there would be less likelihood of polluting the coast. In deep water leaking oil can be sprayed as it comes out of the ship, and the chemical dispersants are less likely to harm the coastline. The harbour master and the Countryside Council for Wales joined the pilots in their call for the ship to be towed away from land. They were overruled by the Government's Joint Response Centre and the commercial salvage crews. Stormy weather made it necessary to evacuate the ship on the following day, and the ship grounded yet again, this time on rocks. Leakage of oil continued until 24 February, when the pumping of oil from the ship finally started. The leakage of oil was estimated as 70 000 tonnes. At the time of writing the damage to wildlife and to the coastline has not been assessed.

A 190 km stretch of coastline was polluted, and sea animals and birds were killed.

FIGURE 6.11D
Milford Haven in South Wales

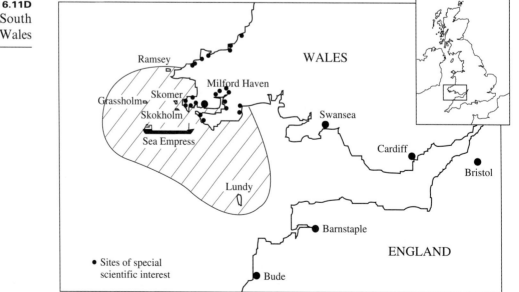

FIGURE 6.11E
A Seabird Killed by an
Oil Spill

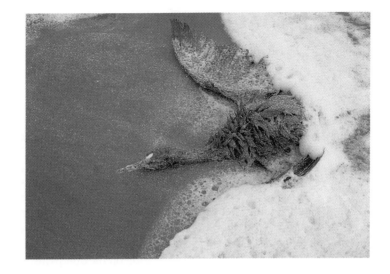

6.12 METHODS OF DEALING WITH OIL SLICKS

DISPERSANTS

Chemical dispersants can be used to break down the slick. They break up the oil into small droplets which are dispersed and diluted by the ocean. In the case of *Exxon Valdez* the dispersants were applied too late after the sea had churned up the oil into an emulsion, a 'chocolate mousse' as one worker described it, which reacted very slowly with the dispersants.

Oil slicks are treated by dispersants ...

SKIMMING

Booms are large fireproof tubes. A boat can place a line of booms around an oil slick to stop the oil spreading. Then skimmers can get to work scooping oil off the surface. In a rough sea this is not easy because oil seeps under the booms. This is what happened in Alaska.

... by skimming off the oil from the surface of the sea ...

BURNING

In some conditions the oil slick can be burnt. Weather conditions must be right; the sea must be calm, the wind must be slight, and the oil must not have started to disperse. The oil is ignited by tossing a torch from a helicopter onto the oil slick.

... by burning ...

SOLIDIFYING

Chemists at British Petroleum have discovered chemicals that can be sprayed onto oil to turn it into a dry, rubbery solid. The oil can then be collected in chunks in nets. If the oil escapes and washes ashore, it can be peeled off the beach like a rubber sheet. The chemicals have been tried out in the laboratory and in experiments at sea. They still have to prove their worth at sea in a real disaster.

... by solidifying and collecting the solid in nets.

CHECKPOINT 6.12

1. (*a*) Mention two ways in which oil from tankers can pollute the ocean.

(*b*) Suggest how this source of pollution can be decreased.

(*c*) Describe some of the damage that an oil slick can do.

2. Discuss methods of dealing with oil slicks.

6.13 RADIOACTIVE POLLUTANTS

Radioactive compounds enter waterways from rain leaching them from the dumps at radium mines and ore-processing plants.

Radioactive compounds get into water supplies by dissolving from the dumps at uranium mines and ore-processing mills. To produce the quantities of uranium needed for nuclear power stations requires the processing of huge quantities of ore which contains only 0.1–0.3% uranium. After uranium compounds have been removed from the ore, the refuse that is left is piled up in dumps or 'tailings'. They contain radioactive thorium, $^{230}_{90}$Th and radium, $^{226}_{88}$Ra. Radium, being in Group 2, is similar in its chemistry to calcium, and if it is ingested into the body it can replace calcium in bones. Thorium is also somewhat similar to calcium and can be deposited in bones. Rain sinking into 'tailings' leaches out radioactive compounds and runs into water supplies. In the USA where uranium ores are processed there are millions of tonnes of tailings piled up, and some river waters have been found to contain twice as much radium as the maximum allowed for human consumption. To reduce leaching from tips, the dumps can be graded steeply so that rain and snow run off quickly and do not penetrate far into the interior. Planting vegetation on the dumps also reduces the dispersion of pollutants by wind and water. See § 8.16 for the disposal of radioactive waste.

6.14 THERMAL POLLUTION

Water is taken from waterways, used for cooling in e.g. power stations, and returned to waterways.

The rise in temperature of the water is called thermal pollution.

Electric power is generated by dynamos driven by steam turbines. When steam leaves the turbine it is liquefied by cooling in a condenser, and the water is returned to the boiler. The condenser is cooled by water pumped from a lake or river near the power station. The cooling water may be warmed by as much as 20 °C before it is returned to its source. The disturbances caused by the rise in temperature of the lake or river is called thermal pollution. It has a number of bad effects.

1. Oxygen is less soluble at the higher temperature. Fish may suffocate and die. The situation is made worse by the speeding up of biochemical processes so that organisms respire more rapidly and need more oxygen.

2. If temperature rises sufficiently high, the biochemical reactions of affected organisms may be so upset that they die immediately.

It decreases the solubility of oxygen ...
... increases the metabolic rate of organisms ...
... speeds up hatching of fish eggs ...

3. The spawning, fertilisation and hatching of fish eggs are geared to water temperature. When these processes are speeded up, the food required by the newly hatched young may not be present. A food shortage will have a disastrous effect on the population of fish.

4. Water which is to be used as a coolant is often chlorinated to prevent the growth of slime in pipes. The chlorine may kill desirable organisms when it is discharged into a river or a lake.

Chlorine in the water may kill organisms.

To avoid thermal pollution, water from a condenser can be cooled by trickling down a tower filled with porous material while a stream of air is blown upwards through it. The heat is transferred to the atmosphere where it may cause less trouble. This may not always be the case in a damp climate such as that of the UK where fogs have sometimes been caused by cooling towers. It works well in a dry climate. An alternative is to use the warm water to cultivate fish or for irrigation to speed the growth of crops.

6.15 MOVEMENT OF POLLUTANTS

It is often difficult to decide whether to classify a pollutant as a pollutant of air or water or land. The case of lead [§ 6.9.2] illustrates this well. Many water pollutants have entered rivers and lakes from air or from land. Water is in motion and transports pollutants so that they may have an effect in a region some distance from the point at which they entered the hydrosphere [see Figure 6.15A].

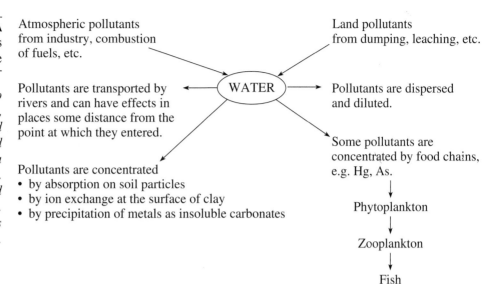

FIGURE 6.15A
Movement of Pollutants
in the Hydrosphere

Pollutants move from air to land, from air to water, from land to water and from water to land They may form part of a food chain. They may be concentrated in soil. Water dilutes pollutants and transports pollutants.

Atmospheric pollutants from industry, combustion of fuels, etc.

Land pollutants from dumping, leaching, etc.

Pollutants are transported by rivers and can have effects in places some distance from the point at which they entered.

WATER

Pollutants are dispersed and diluted.

Pollutants are concentrated
• by absorption on soil particles
• by ion exchange at the surface of clay
• by precipitation of metals as insoluble carbonates

Some pollutants are concentrated by food chains, e.g. Hg, As.

Phytoplankton

Zooplankton

Fish

Mammals

CHECKPOINT 6.15

1. Describe how radioactive compounds can enter the waterways.

2. A number of industries take water from a river, use it for cooling and then return it to the river. Why is this practice described as pollution?

3. Describe what is meant by the concentration of a pollutant through a food chain.

4. Give an example of a substance which can pollute air, water and land.

QUESTIONS ON CHAPTER 6

1. Is the presence of coliform bacteria in water a sign that it is infected with pathogenic organisms?
Why is the coliform test so widely used?

2. Many pollutants in water are converted into harmless products by biodegradation. Why then is their presence in water undesirable?

3. What is an algal bloom? Why is it undesirable?

4. How does mercury get into the aquatic environment? Why is it particularly serious for food production? How can we prevent a build-up of mercury in the environment?

5. Why is heat from industrial processes considered to be a form of pollution?

6. Sewage plants produce quantities of sludge, a thick mud of water and solids which is rich in plant nutrients. Comment on its suitability for use as a fertiliser.

7. Describe the sources and effects of the following pollutants in waterways: cyanides, cadmium compounds, sulphuric acid, aluminium ions.

8. (*a*) Give two sources of lead in the atmosphere.

(*b*) Give one source of lead in drinking water.

(*c*) Suggest another way in which lead could contaminate food.

(*d*) Give one source of lead in soil and dust.

(*e*) Explain how lead can enter the human body.

(*f*) What are the symptoms of lead poisoning?

(g) Mention two steps that have been taken to reduce the quantity of lead in the environment.

9. Phosphates and nitrates may be present at excessive concentrations in rivers.

(*a*) What are the two main sources of these phosphates? Why are phosphates added to the sources you mention?

(*b*) What is the main source of nitrates? Why are nitrates added to the source you mention?

(*c*) Explain how phosphates and nitrates cause eutrophication.

7

THE LITHOSPHERE

7.1 THE EARTH

The lithosphere is the outer layer of Earth's mantle and the crust.

The structure of the Earth is shown in Figure 1.3A. The outer part of the mantle and the crust constitute the **lithosphere** [see § 1.1]. The **crust** is the part of the lithosphere, 5–40 km thick, which humans can penetrate. It consists of rocks. A **rock** is a solid mass of a pure mineral or an aggregate of two or more minerals. A **mineral** is a naturally occurring inorganic solid with a definite crystalline structure and chemical composition.

7.2 ROCKS

7.2.1 IGNEOUS ROCK

Deep inside the Earth are molten rocks called magma. Molten rocks cool and solidify beneath the surface as intrusive igneous rocks or erupt and solidify as extrusive igneous rocks.

At the high temperatures deep beneath the Earth's surface, rocks are in a fluid state. If a crack appears in the crust, molten rock called **magma** erupts as the **lava** of a volcano. When magma cools it crystallises to form **igneous rock**. Examples of igneous rocks are granite, basalt, quartz, feldspar and magnetite. They are hard, porous and of low reactivity. They can be classified as **extrusive igneous rocks** which have cooled rapidly, possibly after a volcanic eruption, to form small crystals and **intrusive igneous rocks** which have cooled more slowly beneath the surface of the Earth and solidified in large crystals.

7.2.2 SEDIMENTARY ROCK

Weathering of rocks produces particles which may collect to form a sediment ...
... which may form sedimentary rock by lithification.

When igneous rocks are exposed to the atmosphere, they very slowly disintegrate through **weathering** due to the action of sun, wind, water, ice and biological and chemical attack. Wind and streams or glaciers pick up particles from weathered rocks and transport them. Eventually the particles are deposited as a bed of **sediment**. As other material is deposited above it, the bed of sediment is compressed until after millions of years the particles join to form a **sedimentary rock**. The process is called **lithification**. Examples of sedimentary rocks are sandstone, shale, limestone (formed from the accumulated shells of marine creatures; see § 4.16) and coal (formed from compressed decayed plants). These rocks are porous, soft and chemically reactive.

7.2.3 METAMORPHIC ROCK

Igneous and sedimentary rocks may be changed by high pressure or high temperature into metamorphic rocks.

Igneous rocks and sedimentary rocks can be changed by high pressure or high temperature into **metamorphic rocks**. Examples are marble, slate and metaquartzite. The conversions between igneous, sedimentary and metamorphic rocks are described by the **rock cycle** [see Figure 7.2A]. The composition of the Earth's crust is: igneous rocks, 65%, sedimentary rocks, 8% and metamorphic rocks, 27%.

FIGURE 7.2A
The Rock Cycle

The rock cycle describes interconversions between rocks.

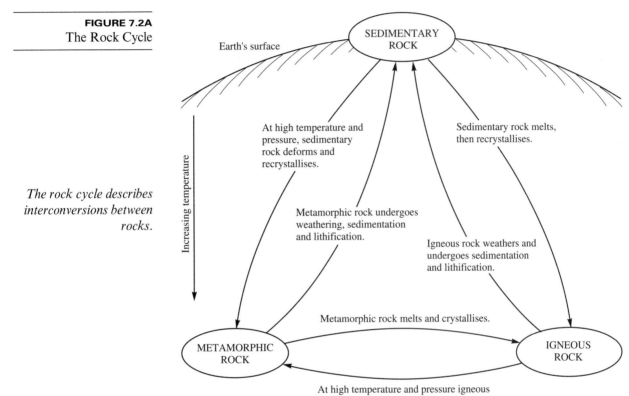

FIGURE 7.2B
Limestone Pavement at Malham Cove

7.3 WEATHERING

Physical weathering of rocks occurs through the action of water, winds, ice, etc.

Weathering of rocks occurs through the action of rain, rivers, glaciers, wind and other agents. This type of weathering – called **physical weathering** – occurs slowly in dry air. Water vastly increases the rate of weathering through its own reactions and through the reactions of the weathering agents which it holds in solution. This type of weathering is called **chemical weathering**. Carbon dioxide, oxygen, organic acids, sulphurous acid, sulphuric acid, nitrous acid and nitric acid play a part. Rainwater attacks minerals by

- hydration of salts such as calcium sulphate, $CaSO_4$
- dehydration of compounds such as hydrated iron(III) oxide , $Fe_2O_3 \cdot nH_2O$
- dissolution of salts such as $CaSO_4 \cdot 2H_2O$ (The enthalpy considerations are discussed in §4.5.)

Chemical weathering takes place through the action of oxygen and acids dissolved in water, especially rainwater, by e.g. dissolution, oxidation, hydrolysis, acid reactions.

- oxidation of compounds such as FeS_2
- hydrolysis of silicates, e.g. Mg_2SiO_4, to soluble compounds
- complexing of e.g. silicates to form soluble compounds
- acid reactions of e.g. dissolved CO_2 and SO_2. The reactions of dissolved carbon dioxide are referred to as **carbonation**, e.g. reaction with calcium carbonate to form soluble calcium hydrogencarbonate.

CHECKPOINT 7.3

1. Describe the origin of igneous, sedimentary and metamorphic rocks.

2. Briefly explain how the following changes take place:

(*a*) sand $\longrightarrow$ sandstone

(*b*) sedimentary rock $\longrightarrow$ metamorphic rock

(*c*) magma $\longrightarrow$ igneous rock

3. (*a*) Why is limestone classified as a sedimentary rock and granite as an igneous rock?

(*b*) What chemical test could you do to distinguish between them?

4. (*a*) Give two examples of physical weathering.

(*b*) How does chemical weathering affect

(i) limestone, (ii) silicates?

7.4 SOIL

Soil is formed by the weathering of rocks and from decaying organic matter.

The most important part of the Earth's crust for living things is **soil**. It is a variable mixture of organic and inorganic matter, including air and water. Soil is formed by the weathering of rocks. The rocks in the Earth's crust have the approximate composition:

O 47%	Si 28%	Al 8%	Fe 5%
Ca 4%	K 3%	Na 3%	Mg 2%

Soil contains air and soil solution – an aqueous solution of many ions.

Other elements present are C, Cl, Mn, P, S. The elements present in rocks are therefore found in soils. The content of carbon and hydrogen is higher when the soil contains organic matter from decaying plant biomass. A typical soil contains 5% organic matter and 95% inorganic matter, including water and air. Peat soils contain up to 95% organic matter, and other soils as little as 1% organic matter. Bacteria, fungi and animals, e.g. earthworms, live in soil. Soil contains air spaces and has a loose texture [see Figure 7.4A]. The water content

Bacteria, fungi and animals live in soil.

of soil is an aqueous solution containing the ions H^+, Ca^{2+}, Mg^{2+}, K^+, Na^+, HCO_3^-, CO_3^{2-}, HSO_4^-, SO_4^{2-}, Cl^-, F^- and others, called **soil solution**.

FIGURE 7.4A
The Structure of Soil

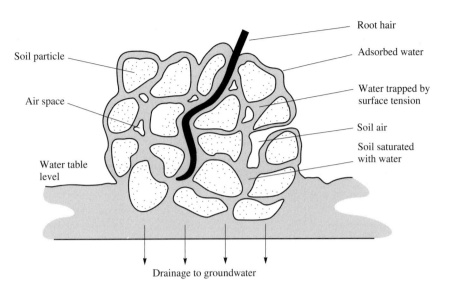

Plants require macronutrients in large amounts and micronutrients in small amounts.

For their growth, plants require the **macronutrients** carbon, hydrogen, oxygen, nitrogen, phosphorus, potassium, calcium, magnesium and sulphur. Of these, carbon, hydrogen and oxygen are available from the air. The rest must be available in soil. Nitrogen, phosphorus and potassium are often added in fertilisers. In addition to these elements which are needed in large amounts, plants need **micronutrients**. These are needed at low concentrations only; at high concentrations they are frequently toxic. Micronutrients are boron, chlorine, copper, iron, manganese, molybdenum, zinc and for some plants sodium, silicon and cobalt.

Layers of soil called horizons can be seen: topsoil, subsoil, weathered rocks and bedrock.

With increasing depth, soils show distinctive layers called **horizons**. Rainwater carries dissolved solids to lower horizons. Bacterial decay of plant biomass produces carbon dioxide and organic acids which are carried to lower horizons where they react with minerals to form new compounds.

FIGURE 7.4B
Soil Horizons

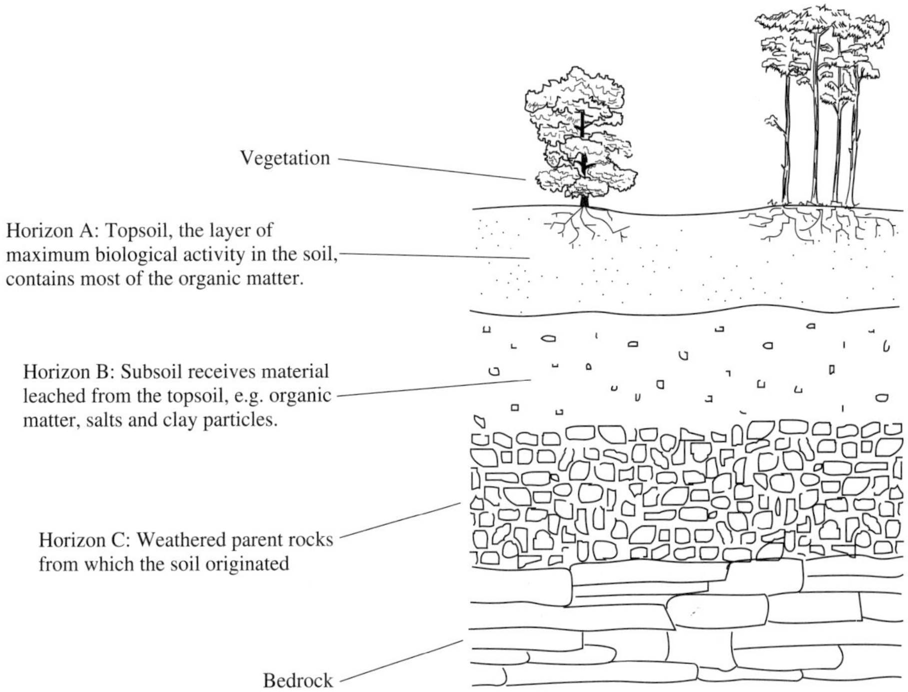

Vegetation

Horizon A: Topsoil, the layer of maximum biological activity in the soil, contains most of the organic matter.

Horizon B: Subsoil receives material leached from the topsoil, e.g. organic matter, salts and clay particles.

Horizon C: Weathered parent rocks from which the soil originated

Bedrock

7.4.1 SOIL TEXTURE CHART

The four major categories of soil particle sizes are:

- gravel, 2–60 mm
- sand, mainly quartz, 0.06–2 mm
- silt, mainly quartz and silicates, 0.006–0.06 mm
- clay, silicates and aluminosilicates, <0.002 mm

Soil components of different particle size are gravel, sand, silt and clays.

The various types of soil are classified by the proportions of sand, silt and clay which they contain, as shown in the soil texture chart in Figure 7.4C.

FIGURE 7.4C
Soil Texture Chart

Use the chart to find out what is the classification of a soil with the composition 40% silt, 40% sand, 20% clay.

Draw lines parallel to the axes as shown.
The lines meet in the area described as loam.

The composition of a soil can be represented by a soil texture chart.

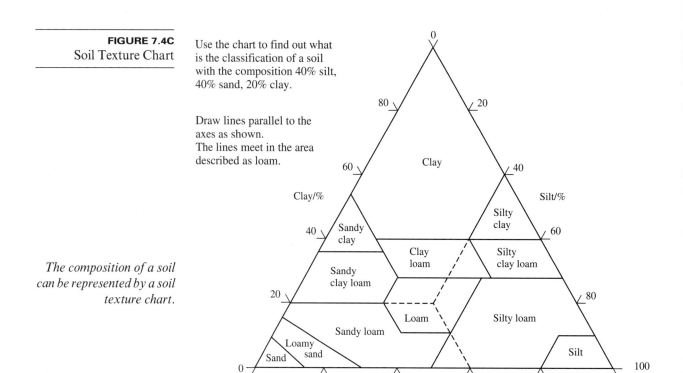

7.4.2 WATER

Clay soils hold the most water, but much of it is hydrogen-bonded to clay and not available to plants.

Clay soil, having the smallest particles, has small spaces between particles, which are small enough to draw water into the clay by capillary action. Clay therefore holds more water than sandy soil, but much of the water is hydrogen-bonded to clay and unavailable to plants. Sandy soil holds less water, but most of the water is available to plants.

7.4.3 AIR

Air is needed by aerobic soil bacteria which decompose organic matter. In the absence of air anaerobic bacteria take over.

Air in soil contains more carbon dioxide than atmospheric air as a result of the decay processes which are occurring. Air is needed by the aerobic bacteria which decompose organic matter. In waterlogged soil, anaerobic bacteria take over the degradation of organic matter and products such as carboxylic acids, methane and hydrogen sulphide are formed.

1. Refer to the soil texture chart in Figure 7.4C. Describe the soils:

(a) **A** = 20% silt, 20% sand, 60% clay

(b) **B** = 20% silt, 60% sand, 20% clay

(c) **C** = 60% silt, 30% sand, 10% clay

2. (a) What is soil and how is it formed from other parts of the lithosphere?

(b) Distinguish between topsoil, subsoil and horizon C.

3. Soils contain sand, silt and clay.

(a) Describe the differences between these components in particle size and in texture.

(b) What is a soil texture chart?

7.5 CLAYS

All clays contain silicates. Most contain aluminium and water. Many contain Na^+, K^+, Mg^{2+}, Ca^{2+}, Fe^{2+}.

Clays are the major inorganic component of most soils. All clays contain silicates, and most contain aluminium and water. Many clays contain sodium, potassium, magnesium, calcium and iron. Clays consist of very fine grains with sheet-like structures. There are three major groups of clays:

- montmorillonite, aluminium hydroxide silicate, $Al_2Si_4O_{10}(OH)_2$
- illite, aluminium potassium hydroxide silicate, $K_{0-2}Al_4Si_{8-6}Al_{0-2}O_{20}(OH)_4$
- kaolinite, aluminium hydroxide silicate, $Al_2Si_2O_5(OH)_4$

Clays bind water and nutrient cations.

Clays are important in holding water and in binding cations such as Na^+, K^+, Ca^{2+}, Mg^{2+}, NH_4^+ . They protect the cations from being leached out by water but keep them available in the soil as plant nutrients.

7.5.1 STRUCTURES OF SILICATES

Silicates are based on the SiO_4^{4-} tetrahedral structural unit.

The layered structures of clays consist of sheets of silicate alternating with sheets of aluminium hydroxide oxide. The unit of silicate structures is SiO_4^{4-} with a tetrahedral distribution of bonds [see Figure 7.5A]. The SiO_4^{4-} anion occurs in minerals such as zircon, $ZrSiO_4$.

FIGURE 7.5A
(a) The SiO_4^{4-} Tetrahedral Ion. The outline of the tetrahedron is shown. The bonds between the Si atom and the O atoms are not shown.
(b) The Tetrahedron

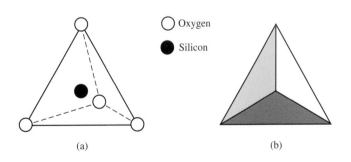

Oxygen
Silicon

(a)   (b)

7.5.2 SILICATE CHAINS

The SiO_4^{4-} units can link up to form long chains of linked tetrahedra. Each SiO_4^{4-} unit shares two of its oxygen atoms with other tetrahedra and has two negatively charged oxygen atoms. These are associated with metal ions which balance the negative charges [see Figure 7.5B]. Some minerals consist of single silicate strands, of formula $(SiO_3)_n^{2n-}$ with associated metal ions.

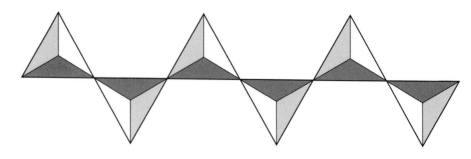

7.5.3 SILICATE SHEETS

Silicate tetrahedra can combine further to form silicates with sheet structures [see Figure 7.5C]. The mineral mica has this type of sheet structure; this is why it is flaky in character.

FIGURE 7.5C
A Silicate Sheet Structure.
Negatively charged
oxygen atoms bond to
layers of metal cations
between silicate sheets,
e.g. Na^+ and Al^{3+}.

*These tetrahedral units can
link up to form chains and
sheet structures . . .
. . . which are negatively
charged and associated
with metal cations . . .*

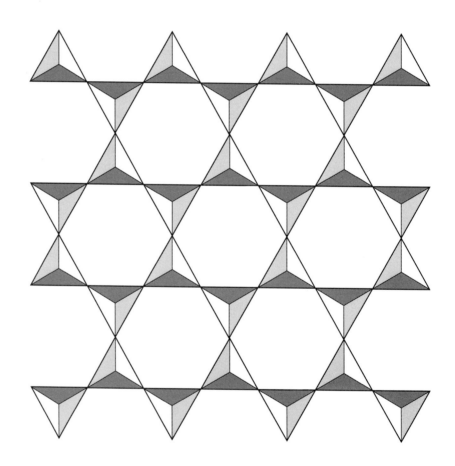

*In clays up to one quarter
of the silicon atoms in the
silicate sheets are replaced
by aluminium atoms . . .*

7.6 CLAY STRUCTURES

*. . . giving the sheet a
negative charge . . .
. . . which is balanced by
associated cations, e.g.
Na^+, Al^{3+}.*

The silicate sheet described in § 7.5.3 is called a **tetrahedral sheet**. Clay minerals contain sheets of this kind. In clays up to one quarter of the silicon atoms in a sheet have been replaced by aluminium atoms. Replacing silicon, with oxidation state + 4, by aluminium, with oxidation state + 3, gives the sheet a negative charge. To balance this negative charge, positively charged ions, e.g. Na^+ and Al^{3+} are held by the sheet. The ions are hydrated and bond water to the clay.

Clays also contain sheets of aluminium hydroxide oxide. The structural unit is octahedral.

Clay minerals also contain sheets of aluminium hydroxide oxide, which consists of aluminium ions surrounded by oxide ions and hydroxide ions in an octahedral arrangement [see Figure 7.6A]. An aluminium hydroxide oxide sheet is called an **octahedral sheet**. Oxygen ions are shared between aluminium ions and also between aluminium ions in an octahedral sheet and silicon atoms in a tetrahedral sheet. The mineral is an **aluminosilicate**.

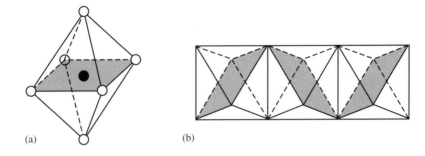

FIGURE 7.6A
(a) The AlO_6 Unit
($\bullet$ = Al, $\circ$ = O; the bonds are not shown),
(b) The Octahedral Sheet of Aluminium Hydroxide Oxide

(a)   (b)

7.6.1 KAOLINITE, A 1 : 1 CLAY

Tetrahedral sheets and octahedral sheets are bonded to form an aluminosilicate.

Kaolinite $Al_2Si_2O_5(OH)_4$, is a clay formed by weathering of feldspar rock. It has a layer structure. Each **unit layer** consists of a tetrahedral silicate sheet and an octahedral aluminium hydroxide oxide sheet which share some oxygen atoms between them. The structure is described as a **two-layer clay** or a **1 : 1 clay** [Figure 7.6B]. The unit layers are joined by hydrogen bonds. These break relatively easily and the clay crumbles readily. Kaolinite is easy to plough and is also used as modelling clay. Oxide ions at the outer surfaces of clay particles attract cations strongly. The cations are hydrated. Water molecules are also hydrogen-bonded to oxide ions on the surface of the silicate tetrahedra, and this water in not available to plants.

FIGURE 7.6B
The Structure of Kaolinite, a 1 : 1 Clay

In a two-layer clay or a 1 : 1 clay, each unit layer consists of a tetrahedral silicate sheet and an octahedral aluminium hydroxide oxide sheet. The unit layers are joined by hydrogen bonds.

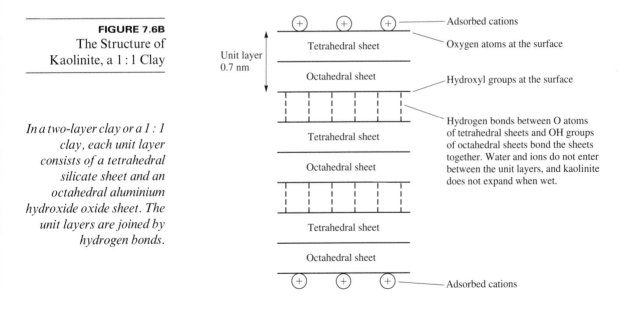

7.6.2 MONTMORILLONITE, A 2 : 1 CLAY

In a **three-layer clay** or a **2 : 1 clay**, a **unit layer** is composed of three sheets. An octahedral sheet shares oxygen atoms with tetrahedral sheets on either side. Examples are montmorillonite and vermiculite, both of which are common in soils. Water molecules are bonded to the surface by hydrogen bonds.

FIGURE 7.6C
The Structure of
Montmorillonite,
a 2 : 1 Clay

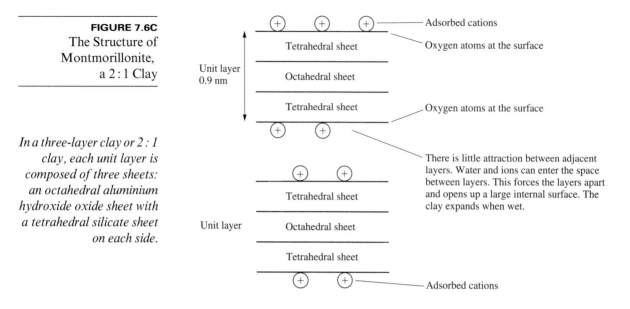

In a three-layer clay or 2 : 1 clay, each unit layer is composed of three sheets: an octahedral aluminium hydroxide oxide sheet with a tetrahedral silicate sheet on each side.

A 2 : 1 clay can absorb more water than a 1 : 1 clay.

Montmorillonite clays may swell by absorbing large quantities of water between unit layers [see Figure 7.6C]. The internal surface is much greater than the external surface; montmorillonite has a surface area of $700–800 \, m^2 \, g^{-1}$ compared with $15 \, m^2 \, g^{-1}$ for kaolinite. The way in which 2 : 1 clays expand when wet makes soils containing them difficult to plough. When the soils dry, the clay contracts and cracks appear in the soil. A 2 : 2 clay is similar to a 2 : 1 clay with an additional octahedral sheet in which some Al atoms are replaced by Fe and Mg atoms.

7.7 HUMUS

Organic matter forms an important part of soil. It provides food for micro-organisms, takes part in ion exchange reactions and improves the structure of soil.

Organic matter in soil is present as only 5% of the mass, but it largely determines the productivity of the soil. It is a source of food for micro-organisms, takes part in ion exchange reactions and influences the physical properties of soil. Organic acids in soil react with minerals to form soluble compounds which release nutrients into the soil. The accumulation of organic matter in soil is influenced by the temperature and the availability of oxygen. In colder climates organic matter does not degrade rapidly and tends to build up. In waterlogged soils decaying vegetation lacks oxygen and organic matter may build up to as high as 90% .

Humus is a part of the organic matter, the residue left when bacteria and fungi degrade plant material.

The most important component of the organic material is **humus**, a material which is insoluble in water and biodegrades slowly. It is the residue left when bacteria and fungi degrade plant material. Humus consists of a number of **humic substances**: an insoluble fraction called **humin** and two base-soluble fractions called **humic acid** and **fulvic acid**.

Humic substances have a big influence on the properties of soils.

- They increase the capacity of soil for holding water.

- They bind metal cations through their carboxyl groups and phenolic hydroxyl groups. Insoluble humic substances may accumulate large quantities of metal cations, thus holding micronutrients in the soil.

It holds water in the soil, bonds to cations, has a buffering action and bonds to organic and inorganic compounds in soil.

- Humic substances exert a buffering action on soil.

- They help soil particles to cling together by forming a surface coating on mineral grains.

● Humic substances increase the adsorption of organic compounds by soil. They have a strong affinity for organic compounds of low solubility, e.g. DDT and some herbicides.

● Humic substances may interact with inorganic components of soil by the formation of complexes. In many soils 50% of the organic content is complexed with clay.

If humus is present in water during chlorination, trihalomethanes, which are suspected carcinogens, are formed.

Humic substances have been a focus of attention since 1970 when trihalomethanes, e.g. $CHCl_3$, $CHClBr_2$, were found in drinking water. These compounds are suspected of being carcinogenic. Humic substances react with chlorine to produce trihalomethanes, so it is believed that trihalomethanes are formed when the municipal water supply is chlorinated [see § 5.4]. The contamination can be reduced by removing as much humus as possible before chlorination.

CHECKPOINT 7.7

1. Silicate clays consist of tetrahedral sheets and octahedral sheets.

(*a*) Describe the structural unit of (i) a tetrahedral sheet, (ii) an octahedral sheet.

(*b*) By means of a sketch, explain how the arrangement of sheets differs in a 1 : 1 clay and a 2 : 1 clay. Say what difference in properties between the two clays results.

7.8 CATION EXCHANGE

Magnesium can replace silicon and aluminium in clays, giving the clay a net negative charge which makes it bond to cations.

Clay minerals have some of their silicon(IV) and aluminium(III) replaced by metal ions of a similar size but lower charge, e.g. Mg^{2+}. The clay attains a net negative charge. This is balanced by cations held at the surfaces of the clay layers [Figures 7.6B and C]. The cations are hydrated and give clay its sticky feel. These cations are not part of the crystalline structure, and they are exchangeable for other cations in the aqueous solution in soil. The amount of exchangeable cations in a clay is called the **cation-exchange capacity**. It is expressed as the amount of monovalent cations per kg of clay (that is, amount of Na^+/mol + $\frac{1}{2}$ amount of Mg^{2+}/mol + $\frac{1}{3}$ amount Al^{3+}/mol, etc). Typical values $(mol\,kg^{-1})$ are kaolinite 0.3, montmorillonite 1.0, vermiculite 1.5, humus 1.5–3.0, soil 0.02–0.6. The adsorbed ions are important in plant nutrition. If plant roots take potassium ions out of the soil, more potassium ions can be released from adsorption to clay.

The cations can be exchanged. The amount of exchangeable cations in a clay is the cation-exchange capacity.

The cation exchange capacity is of major importance in the ability of a soil to supply nutrients to plants. Between clay and the water present in soil, exchanges of the following type take place:

$$Ca^{2+}(clay) + 2H_2O(l) \rightleftharpoons 2H^+(clay) + Ca^{2+}(aq) + 2OH^-(aq)$$

The adsorbed ions are important in nutrition.

Metal cations and ammonium cations accumulate in soil solution. From the solution they can be either absorbed by plants or washed out by rain and lost. If the cations were not held by clay or humus, they would be free in the soil solution and liable to be easily washed out by rain.

Organic acids in humic substances act as cation exchangers.

Humic substances contain organic acids which can act as cation exchangers. These weak acids are partially dissociated:

$$RCO_2H(soil) \rightleftharpoons RCO_2^-(soil) + H^+(aq)$$

Exchange takes place between clay and soil solution. When ions are taken up by plants they can be replaced by desorption from clay.

The anions can bind metal cations.

$$RCO_2^-(soil) + M^+(aq) \rightleftharpoons RCO_2^-M^+(soil)$$

Ion exchange is reversible. An exchange may occur between calcium ions on clay and ammonium ions in soil solution:

$$Ca^{2+}(clay) + 2NH_4^+(aq) \rightleftharpoons Ca^{2+}(aq) + 2NH_4^+(clay)$$

The exchange is reversible, the direction depending on the concentrations of the ions and the relative strengths of their bonds to the clay. Plant roots draw nutrients from soil solution. If plant roots are withdrawing ammonium ions from soil solution, the equilibrium moves from right to left and ammonium ions are released by the clay.

Ion exchange is reversible.

Ammonium ions are more strongly bound to soil than nitrate ions. When nitrate fertilisers are added, some of the nitrate is taken in by plants and the excess is rapidly leached from the soil. When ammonium fertilisers are added, ammonium ions are bound to the soil and gradually oxidised to nitrate ions:

$$NH_4^+ + 8O_2(g) \longrightarrow 4NO_3^-(aq) + 8H^+(aq) + 4H_2O(l)$$

Ammonium ions are more strongly bound to soil than nitrate ions.

The reaction increases the acidity of the soil. If the soil acidity becomes too high the reverse reaction, the reduction of nitrate to ammonium ion is promoted.

7.9 SOIL pH

Hydrogen ions enter the soil in the rain and from the activities of micro-organisms. There is another reason why, in the natural course of events, soil becomes gradually more acidic. Cation exchange takes place between clay and soil solution [see § 7.8]:

Soil becomes gradually more acidic in the natural course of events. Cations are removed from soil solution by plants and replaced by hydrogen ions.

$$Ca^{2+}(clay) + 2H_2O(l) \rightleftharpoons Ca^{2+}(aq) + 2H^+(clay) + 2OH^-(aq)$$

As cations are removed by plants or by leaching, the equilibrium moves from left to right. The hydrogen ions formed are not taken up by plants; they accumulate and the pH falls. There is therefore a natural tendency for soil to become more acidic as a result of the removal of nutrients for plant growth.

The oxidation of minerals such as pyrite, FeS_2, in soil causes a rise in acidity.

$$2FeS_2(s) + 7O_2(g) + 2H_2O\,(l) \longrightarrow 2Fe^{2+}(aq) + 4H^+(aq) + 4SO_4^{2-}(aq)$$

Minerals such as sulphides are oxidised to form acids.

In addition FeS_2 reacts with acid to form hydrogen sulphide which is toxic to plant roots. These reactions happen especially in marshland reclaimed for agriculture. Marshlands are formed from marine sediments which often contain pyrite. Mine spoils (minerals left over from mining) also contain pyrite and acidify soil in the same way.

7.9.1 THE EFFECT OF pH ON ION EXCHANGE

The cation exchange capacity of a soil depends on the pH. Variations in soil pH therefore control the availability of nutrients to plants.

At low pH, hydrogen ions displace other cations from adsorption on soil . . .

1. At low pH, hydrogen ions displace exchangeable cations from soils.

$$Ca^{2+}(clay) + 2H^+(aq) \rightleftharpoons Ca^{2+}(aq) + 2H^+(clay)$$

... thus reducing the nutritive value of soil.

In this way the increased acidity of soil solution reduces the ability of the soil to hold exchangeable cations (other than H^+) and reduces the nutritive value of the soil. The actual hydrogen ion concentration in the soil solution is called the **active acidity**. The concentration of hydrogen ion held on exchange sites is called the **reserve acidity**. When the active acidity of a soil increases, metal ions are leached away and replaced by hydrogen ions so that the reserve acidity increases also.

2. At low pH hydrogen ions in soil solution react with hydroxide groups at the surface of silicate sheets:

They also react with the hydroxide groups at the surface of silicate sheets.

$$Clay—OH(s) + H^+(aq) \rightleftharpoons Clay—OH_2^+(s)$$

A positively charged surface is created, and it repels cations, which move into the soil solution. The surface can hold anions, e.g. phosphate ions. If phosphate ions are taken from the soil by plants, the concentration in soil solution can be replenished by desorption from the surface of clay. See also § 7.10.

At high pH, hydroxide ions in soil solution react with hydroxide groups at the surface of silicate sheets. The result is a negative surface which binds cations. This makes Ca^{2+}, Mg^{2+} and K^+ ions accumulate in soil solution.

3. At high pH, hydroxide ions in soil solution react with hydroxide groups at the surface of silicate sheets:

$$Clay—OH(s) + OH^-(aq) \rightleftharpoons Clay—O^-(s) + H_2O(l)$$

The reaction leaves the surface with a negative charge which can bind exchangeable cations. At very high pH, however, calcium and magnesium ions are removed from solution by precipitation as insoluble hydroxides and carbonates.

7.9.2 EFFECT OF pH ON Al^{3+} AND Fe^{3+} CONCENTRATIONS

Weathering releases ions into solution. The hydrolysis of the aqua ions $[Al(H_2O)_6]^{3+}$ and $[Fe(H_2O)_6]^{3+}$ forms hydrogen ions, which maintain the acidity of the soil and cause further weathering.

Silicate clays contain aluminium ions and iron ions. At low pH the weathering of clay minerals accelerates and releases the hydrated ions into soil solution.

$$Al_2Si_2O_5(OH)_4(s) + 6H^+(aq) \longrightarrow 2Al^{3+}(aq) + 2Si(OH)_4(s) + H_2O(l)$$
Kaolinite

Both $[Al(H_2O)_6]^{3+}(aq)$ ions and $[Fe(H_2O)_6]^{3+}(aq)$ ions are small in size and highly charged and undergo hydrolysis to release hydrogen ions.

$$[Al(H_2O)_6]^{3+}(aq) + H_2O(l) \rightleftharpoons [Al(OH)(H_2O)_5]^{2+}(aq) + H_3O^+(aq)$$

$$[Al(OH)(H_2O)_5]^{2+}(aq) + H_2O(l) \rightleftharpoons [Al(OH)_2(H_2O)_4]^+(aq) + H_3O^+(aq)$$

High concentrations of aluminium ions are toxic to crops.

Hydrogen ions consumed in weathering the clay are replaced through hydrolysis, maintaining the acidity of the soil and causing the release of more aluminium ions and iron ions into soil solution. Being highly charged, aluminium ions and iron ions bond strongly to exchange sites. High concentrations of Al^{3+} ions in soil solution are toxic to crops, and at pH values below 4 the concentration of Al^{3+} ions in solution is high enough to limit plant growth and to reduce beneficial bacteriological action in the soil.

7.9.3 EFFECT OF pH ON SOLUBILITY OF HEAVY METAL IONS

Low pH favours the formation of more soluble compounds of heavy metals which are more readily leached from the soil.

The lower oxidation states of metals, e.g. Fe(II), Mn(II) tend to form more soluble compounds than those of higher oxidation states. The rate of oxidation to higher oxidation states and insoluble compounds, e.g. Fe_2O_3 and MnO_2, increases with pH. Acidic soils favour the lower oxidation states and the formation of soluble salts which can be leached out of the soil.

7.10 CONTROL OF pH

There are natural buffers present in soil. Humic substances contain organic acids which act as buffers.

$$RCO_2H(aq) + H_2O(l) \rightleftharpoons RCO_2^-(aq) + H_3O^+(aq)$$

When the pH is low, the above equilibrium moves from right to left.

When the pH is high, the equilibrium below moves from left to right.

$$RCO_2H(aq) + OH^-(aq) \rightleftharpoons RCO_2^-(aq) + H_2O(aq)$$

Natural buffers present in soil include organic acids in humus and hydrogencarbonate ions.

Hydrogencarbonate ions are formed when carbon dioxide dissolved in water reacts with carbonates:

$$CO_2(aq) + H_2O(l) + CaCO_3(s) \rightleftharpoons Ca(HCO_3)_2(aq)$$

Hydrogencarbonates have a buffering action.

$$HCO_3^-(aq) + H_3O^+(aq) \rightleftharpoons H_2CO_3(aq) + H_2O(l)$$

$$HCO_3^-(aq) + OH^-(aq) \rightleftharpoons CO_3^{2-}(aq) + H_2O(l)$$

Calcium oxide and calcium carbonate are spread on soil to neutralise excess acidity.

In order to increase the pH and avoid accumulation of Al^{3+} ions in soil solution, ground limestone or chalk ($CaCO_3$) or calcium oxide (quicklime, CaO) can be spread on the soil surface. Different clay minerals need different applications of base because they have different buffering capacities [see Figure 7.10A].

FIGURE 7.10A
Buffering Capacities of
Some Clay Minerals

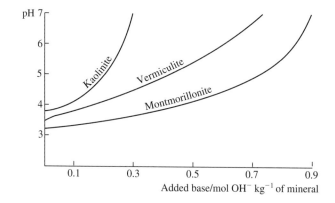

In raising the pH of soil it is necessary to neutralise hydrogen ion in soil solution (the active acidity). For a more permanent effect, it is necessary to replace the hydrogen ions at the exchange sites (the reserve acidity), usually by calcium ions, Ca^{2+}, although magnesium ions, Mg^{2+}, or potassium ions, K^+, would also be effective. The addition to soil of calcium carbonate or calcium hydroxide reduces the reserve acidity of the soil and has an effect which lasts for 2–10 years. The presence of a fair quantity of Group 1 and Group 2 metal ions in a soil indicates that it is not highly acidic. If the soil is too alkaline – a less common problem – it can be treated with humus or with sulphur which is slowly oxidised to sulphuric acid, or with an acidic fertiliser such as ammonium sulphate.

A soil which is too alkaline is treated with humus or with sulphur.

In a cold climate, such as that of Norway and Sweden in the winter, acid snow accumulates. When it melts suddenly in the spring, there is a rapid release into the soil of a large volume of acidic water. This may tax the ability of the lime in the soil to neutralise it.

1. Explain why ammonium sulphate lowers the pH of soil solution.

2. Why is it difficult for plants to obtain all the nutrients they need in a soil with a high proportion of cationic exchange sites occupied by hydrogen?

3. Explain how soil may become acidic because of the natural process of supplying nutrients for plant growth.

4. Refer to Figure 7.10A which shows the buffering capacities of different clays. Suggest why montmorillonite has a better buffering capacity than kaolinite.

5. What role does humus play in (a) the regulation of soil pH and (b) the retention of cations by soil?

6. Explain how silicate clays can have a permanent cation exchange capacity and a pH-dependent cation exchange capacity.

7. (a) Explain why free aluminium ions are more available in soil at low than at high pH.

(b) Explain why the presence of aluminium ions in soil lowers its pH still further.

(c) Why is it difficult to restore the pH to a higher value?

8. A sample of soil is treated with a $1 \, mol \, dm^{-3}$ solution of K^+ ions, followed by a $1 \, mol \, dm^{-3}$ solution of Ca^{2+} ions, followed by a $1 \, mol \, dm^{-3}$ solution of Mg^{2+} ions. Explain what changes take place at the cation exchange sites in the soil as a result of each treatment.

7.11 AVAILABILITY OF IONS

The availability of different ions as plant nutrients depends on the pH.

The growth of plants depends on the availability of nutrients, and the presence of ions in solution depends on the pH.

1. Iron, manganese and aluminium are more available in acidic solutions; at pH > 6 Fe^{3+} is precipitated as $Fe(OH)_3(s)$. In some cases low pH conditions release excessive and toxic concentrations of the ions.

Some ions are leached away at low pH.

2. Calcium and magnesium are removed from soil solution at pH > 8.5 to form insoluble carbonates. At low pH they are leached away [see §7.9].

Some ions are precipitated as insoluble compounds at high pH.

3. Phosphate ion concentration is reduced at low pH because of the insolubility of iron(III) phosphate and aluminium phosphate, and at high pH because of the insolubility of calcium phosphate.

4. Sulphate ion concentration is reduced at pH < 6 to form insoluble sulphates.

At low pH, nitrate ion is reduced to ammonium ion.

5. Nitrogen is present in soils as NH_4^+ at pH < 5.5 and as NO_3^- at pH > 5.5. In an acidic soil, nitrate ions may be reduced to ammonium ions.

$$NO_3^-(aq) + 10H^+(aq) + 8e^- \rightleftharpoons NH_4^+(aq) + 3H_2O(l)$$

At high pH ammonium ion forms ammonia.

Plants take in nitrates but not ammonium salts; therefore a low soil pH reduces the availability of nitrogen for plant nutrition. At very high pH ammonium ions are converted into ammonia which vaporises from the soil. An advantage of keeping the soil well aerated is that the oxygen in it will keep nitrate in its oxidised state. Denitrification occurs when the oxygen content of soil is low and anaerobic bacteria reduce nitrate ions. It is the natural mechanism by which fixed nitrogen is returned to the atmosphere [see the nitrogen cycle, §2.6]. Below pH 5.5 nitrogen fixation in the root nodules of legumes ceases and the decomposition of humus by nitrifying bacteria is greatly slowed down.

The best range of pH overall is 6.0–6.5.

6. The best range of pH overall is pH 6.0–6.5.

FIGURE 7.11A
pH Ranges for the
Availability of Plant
Nutrients in Soil

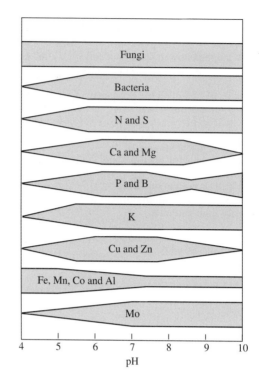

7.12 SOIL MANAGEMENT

There are natural mechanisms for assisting soil to conserve its nutritive value.

Soil can renew itself and conserve its nutritive value without outside intervention. Rain can be absorbed in the pores between soil particles, keeping erosion to a minimum. Wind erosion is slowed by aggregation of soil particles and the holding power of plant roots. Soil bacteria, fungi and animals can remove and detoxify a variety of harmful substances that have been deposited from the air and from water. The process is assisted by catalysis at the surfaces of colloidal clays. Air pollutants, e.g. carbon monoxide, insecticides, herbicides, plant and animal remains can all be assimilated in moderate quantities. Inspite of all these safeguards, soils can become degraded through human activity.

When crops are harvested, they do not return nutrients to the soil, and fertilisers must be added.

When crops are harvested instead of being allowed to decay they do not return nutrients to the soil, and fertilisers must be added. They replace lost nutrients but they do not restore the structure and porosity that organic matter contributes to the soil. It is wise for the farmer to leave enough crop residues to maintain the organic matter at a level high enough to nourish crops and to prevent erosion.

Fresh plant biomass has a C : N ratio of 100 : 1. On complete humification the C:N ratio is 10 : 1. If the C : N ratio of organic matter in soil is too high, nitrogenous fertilisers are added.

Fresh plant biomass has a C : N ratio of 100 : 1. Organic matter is decomposed in soil by bacteria and fungi to form humus with a C : N ratio of 10 : 1. If the C : N ratio of the organic matter in the soil is too high, nitrogen may be the limiting factor in the growth of organisms which decompose organic matter and recycle nutrients. If straw (C : N = 80 : 1) is ploughed into the soil, a nitrogenous fertiliser is usually applied to lower the C : N ratio. Composting can be used to reduce the C : N ratio: storing organic matter in a compost pile with moisture and air allows carbon dioxide and water to escape, while nitrogen is retained in the amino acids and proteins of micro-organisms. Adding fertiliser to compost increases the population of micro-organisms and speeds composting. The addition of fertilisers to a soil which contains little organic matter is not helpful. There is no organic matter to retain nitrogen, and large amounts of aluminium oxide and iron(III) oxide in weathered soils tend to precipitate phosphate as aluminium phosphate and iron(III) phosphate.

Adding fertilisers to a soil which contains little organic matter does not help.

7.12.1 METHANE

In the presence of oxygen, bacteria obtain the energy which they need for their life processes by oxidising organic matter to carbon dioxide and water.

$$(CH_2O)(s) + O_2(g) \longrightarrow CO_2(g) + H_2O(l)$$

When the oxygen content of soil is low anaerobic microbes decompose organic matter with the formation of methane ...

When the oxygen content of the soil is low, anaerobic microbes take over and partial decomposition takes place. This happens in soil and in water. The production of methane in oxygen-deficient sediments is favoured by high levels of organic matter and low nitrate and sulphate levels. This reaction plays a key role in the carbon cycle [§ 2.8]. When organic matter (CH_2O) is degraded by anaerobic microbes the reaction is:

$$2(CH_2O)(s) \longrightarrow CH_4(g) + CO_2(g)$$

... and hydrogen sulphide.

Hydrogen sulphide is another product of the anaerobic decay of organic matter.

CHECKPOINT 7.12

1. A farmer has a choice between farmyard manure and synthetic fertilisers.

(a) Give two reasons why the farmer might choose manure.

(b) Why is it not the universal choice?

2. (a) What is the $C:N$ ratio of organic matter?

(b) How does ploughing in straw alter the $C:N$ ratio of soil?

(c) How does composting reduce the $C:N$ ratio?

(d) What is the advantage of applying nitrogenous fertiliser at the same time as organic matter?

QUESTIONS ON CHAPTER 7

1. (a) List the three main groups of rocks in the Earth's crust. Briefly describe how each type is formed.

2. Outline what happens in:

(a) physical weathering

(b) chemical weathering.

3. (a) What visible difference would you observe between an extrusive igneous rock and an intrusive igneous rock?

(b) How could you distinguish limestone from an igneous rock?

4. Nitrate ions are formed in soil by the oxidation of ammonium ions. The half-equation for the reaction is

$$NH_4^+(aq) + 3H_2O(l) \longrightarrow NO_3^-(aq) + 10H^+(aq) + 8e^-$$

(a) State how the reaction affects soil acidity.

(b) Explain why nitrate ions are more easily leached from soils than ammonium ions.

(c) Suggest why it is not a good idea to apply nitrate fertilisers in the autumn.

(d) Explain how nitrate ions contribute to eutrophication of water.

5. (a) Suggest one reason why a soil might become anaerobic.

(b) State two reduction reactions that could occur in soil under anaerobic conditions, and explain why each would be disadvantageous for plant growth.

6. (a) What is humus?

(b) What is the $C:N$ ratio of humus?

(c) What role does humus play in the retention of cations by soil?

7. (a) Distinguish between *active acidity* and *reserve acidity* of a soil.

(b) Explain the effect of liming on these two types of soil acidity.

(c) Suggest a pH which is suitable for general plant growth. Explain why higher and lower pH values are less suitable.

8. (a) Sketch the shape of the silicate unit SiO_4^{4-}.

(b) Explain how silicate units are combined in kaolinite.

(c) In some clays some of the silicon is substituted by aluminium.

 (i) Explain why this substitution causes the clay particles to carry a negative charge.

 (ii) Name two ions that are commonly adsorbed on such clay particles.

(iii) Describe one consequence of such adsorption.

(d) A clay soil and a sandy soil have a pH of 6.0. Explain why more lime has to be added to the clay soil than to the sandy soil to neutralise the acidity.

9. (*a*) Water is one of the agents that weathers rocks. They are weakened by the freezing and thawing of water in cracks in the rock. Explain why freezing and thawing weakens rocks, and why water is ideally suited to do this.

(*b*) Removal of minerals through the action of rain, which is a dilute solution of carbon dioxide, is another means of weathering. Name one component of rocks that can be removed in this way. Give an equation for its reaction with aqueous carbon dioxide.

(*c*) Hydration of metal ions by water is another agent for weathering rocks. What considerations of (i) enthalpy, (ii) entropy favour the dissolution of a mineral from a rock in this way?

10. The cation exchange capacity (CEC) of a soil is measured at two different pH values.

pH	CEC of organic material	CEC of silicates
3.0	29	35
7.5	155	58

(*a*) Explain why the CEC of (i) the organic material, (ii) the silicates increases as pH rises.

(*b*) Suggest why the CEC value of the silicates is greater than that of the organic material at low pH and less at higher pH.

(*c*) To maintain a high CEC it might be a good idea to keep the soil at pH 7.5. There is a drawback; one essential ion could be lost from the soil. What is this ion? How could it be removed from the soil?

11. A soil which is rich in humus is irrigated with a solution of a fertiliser containing potassium ion and ammonium ion. The solution becomes depleted of these nutrients as it passes through the soil. How has this happened?

12. Ammonia is a popular fertiliser. Most plants assimilate nitrogen preferentially as nitrate ion.

(*a*) What essential role do bacteria play when ammonia is used as a fertiliser?

(*b*) What problem arises in waterlogged soils?

8

LAND POLLUTION

8.1 POLLUTANTS TRAVEL

There is an exchange of pollutants between land, water and air.

Many of the substances mentioned here as land pollutants are also water pollutants because they are leached out of the soil into waterways. There is a constant exchange of substances between land and air. Pollutants in the air, e.g. soot and acid rain, fall on to the land. Substances formed in soil, e.g. hydrogen sulphide and methane, pass into the atmosphere. Lead compounds pollute the air, for example when they are part of the exhaust gases of vehicle engines. They also pollute land when they are are deposited on the soil alongside the roads. From there they can be washed into waterways and into groundwater. Pesticides are applied to land where they may remain as pollutants. In addition, there is a possibility of pesticides entering water directly from applications such as mosquito control and indirectly from drainage of agricultural land.

8.2 SOIL POLLUTION

Soil receives large quantities of pollutants . . .

Soil is the medium that produces most of the food required by living things. Good soil is a valuable asset. Soil receives large quantities of pollutants:

1. Particulate matter from power plant chimneys and other sources falls on to the land [§ 3.1].

. . . e.g. soot, sulphur dioxide and oxides of nitrogen . . .

2. Sulphur dioxide emitted by sulphur-containing fuels ends up on soil as sulphates [§ 3.6].

. . . carbon monoxide, which is converted into harmless products . . .

3. Atmospheric nitrogen oxides are converted into nitrates and eventually fall on to soil in the rain [§ 3.7].

4. Soil absorbs carbon monoxide and microbial action converts it into carbon dioxide and possibly into biomass. Carbon monoxide emissions in urban areas do not have this sink available [§ 3.3].

. . . lead compounds . . .
. . . leachate from landfill sites . . .
. . . and degradable hazardous wastes.

5. Particles of lead compounds from vehicle exhausts fall on to soil alongside roads [§ 3.2].

6. Soil receives the liquid leachate that flows out of landfill sites [§ 8.7].

7. In some cases soil is used as a depository for degradable hazardous wastes which are worked into the soil and left for soil microbial processes to bring about degradation.

Sewage and sewage sludge may be spread on soil. Fertilisers and pesticides applied to soil may pollute the atmosphere and the waterways.

8. Sewage and sewage sludge may be applied to the soil [§ 5.6].

9. Substances applied to soil as fertilisers and pesticides often contribute to the pollution of water and air [§§ 6.4, 8.3, 8.4].

8.3 PESTICIDES

Pesticides are organic and inorganic compounds that are used to improve the human environment by controlling undesirable living things. They include insecticides, herbicides, plant growth regulators, fungicides, bactericides and algicides. Herbicides account for two-thirds of agricultural pesticides. Insecticides and fungicides are the most important with respect to human exposure because they are applied shortly before harvesting or even after harvesting crops.

There are persistent pesticides which are not degraded by chemical and biological activity. They are **bioaccumulative**, that is they are retained within the body of the organism which ingests them. They are concentrated at each level of the food chain.; For example, the insecticide DDT is applied to an area at a concentration of less than one part per billion. When it is ingested by microcopic organisms the concentration may increase up to a thousandfold. As these organisms are ingested by higher forms of life, algae, fish, birds or humans, the concentration at the end of the food chain may reach levels of thousands or millions of parts per billion. Figure 8.3A shows what happened when DDT was used to spray Clear Lake in California to get rid of mosquitoes.

FIGURE 8.3A
A Food Chain in Clear Lake, California

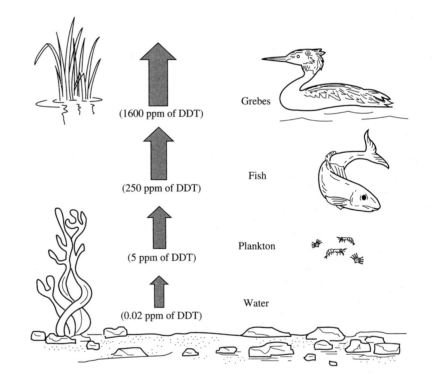

(1600 ppm of DDT)   Grebes

(250 ppm of DDT)   Fish

(5 ppm of DDT)   Plankton

(0.02 ppm of DDT)   Water

8.4 INSECTICIDES

8.4.1 FUNCTION

Insecticides function either

- as stomach poisons, which must be ingested
- as contact poisons which act through the skin
- as fumigants, which must be inhaled.

Chemically speaking, insecticides are of the following kinds:

- inorganic compounds
- organochlorine compounds, e.g. DDT
- organophosphorus compounds, e.g. parathion
- carbamates, which possess the group $>N\!-\!CO_2\!-\!$
- plant extracts, e.g. pyrethrins and pyrethroids

8.4.2 ORGANOCHLORINE INSECTICIDES

Organochlorine insecticides are exemplified by DDT.

The substances known as **organochlorine insecticides** are organic compounds containing chlorine. The most famous is DDT.

8.4.3 DDT

A research student called Othmar Geidler made DDT in 1874. Sixty years later, another chemist called Paul Mueller repeated the synthesis so that he could try out DDT in his work on insecticides. He found that it was extremely poisonous to houseflies and other insects.

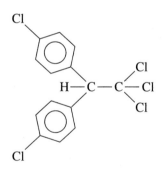

1,1,1-Trichloro-2,2-bis(4-chlorophenyl)ethane
(The letters DDT come from its former
name, dichlorodiphenyltrichloroethane.)

DDT, a chloro-compound with insecticidal properties, was used against mosquitoes and lice in the Second World War.

This work was done in the 1930s, and when the Second World War started in 1939, chemists had found more uses for DDT. During a war, in addition to those killed in action, many people die of disease through lack of medical supplies, shortage of water, overcrowding and poor sanitation. The use of DDT in the Second World War helped to alleviate some of this misery. When the Allies landed in islands in the Pacific, they faced the danger of malaria as well as enemy forces. By spraying DDT from aeroplanes, they were able to wipe out the mosquito population, and remove the source of malaria. Dusting with DDT kept the troops free from the body lice which had plagued soldiers in earlier wars. After the Allies landed in Italy and occupied Naples, an epidemic of typhus broke out. To kill the lice which carry the disease, the whole area was sprayed with DDT and the population dusted themselves with DDT. Within days, the epidemic was over. In 1948, Mueller was awarded the Nobel prize, for discovering the life-saving properties of DDT.

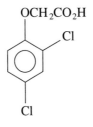

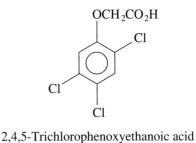

2,4-Dichlorophenoxyethanoic acid
2,4-D

2,4,5-Trichlorophenoxyethanoic acid
2,4,5-T

DDT and related compounds were used as insecticides and herbicides by farmers.

After the war, farmers welcomed DDT-related compounds to replace the non-selective insecticides and herbicides which they had been using. DDT killed insects but not farm animals. The herbicides 2,4-D and 2,4,5-T killed weeds and left grass to grow. Dieldrin and aldrin were found to be even more potent than DDT:

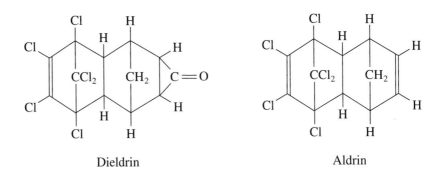

Dieldrin   Aldrin

As early as 1946, however, it began to appear that the new chloro-compounds were not a perfect solution to the insect problem. Some species of housefly soon became

Some insects soon became resistant.

resistant to DDT. In the USA, the cotton farmers had been delighted with the way DDT had attacked the cotton boll weevil, but, by 1960, they were having to spray more and more frequently with higher and higher doses as the weevil became resistant to the insecticide.

These compounds are very stable, and remain in a treated area to be eaten by micro-organisms, fish, birds and small animals.

In 1962, Rachel Carson, in her book *The Silent Spring*, called for a halt to the widespread and indiscriminate spraying of insecticides and herbicides. These compounds are so stable that they persist for a long time in areas where they have been sprayed. If they are eaten by birds and animals, they cannot be excreted because they are insoluble in water, but they can be stored in the body because they are soluble in fat. In the spring of 1956, large numbers of birds were found dead in cereal-growing regions and were found to have a high content of dieldrin and aldrin, with which the cereals had been treated. Scientists began to investigate the spreading of insecticides and herbicides. They found DDT in penguins in the Antarctic, where the spray had never been used. They reasoned that if DDT is used in a malarial region, it will be taken up by small organisms. If these are eaten by a fish and the fish is eaten by a bird, the bird may carry DDT for hundreds of miles. DDT can concentrate up a food chain. If the content of DDT in seawater is 1×10^{-6} ppm, plankton can have a content of 3×10^{-4} ppm, and fish eating the plankton may contain 0.5 ppm. Birds of prey concentrate DDT further until they may contain 10 ppm. It is possible that fish and birds which have high DDT contents may be eaten by human beings. If DDT is so toxic to insects, can it be completely harmless to human beings? In 1964, the Advisory Committee on Poisonous Substances used in Agriculture and Food Storage recommended that dieldrin and aldrin should be used only when drastic measures are

The use of DDT has declined.

needed. Restrictions have been placed on the use of DDT, and consumption of DDT has fallen to about half of what it was in 1964.

FIGURE 8.4B
A Model of a DDT
Molecule

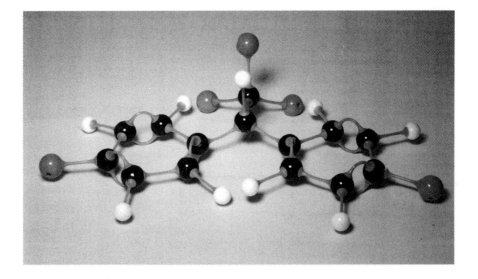

Toxaphene and **lindane** are organochlorine insecticides. For some time methoxy-chlor was a popular substitute for DDT, having a low toxicity to mammals and biodegrading in a reasonable time. Chlordane, aldrin, dieldrin, endrin and heptachlor were used as insecticides and all are now banned because of the long time for which they persist in soil and because of suspicions that they may be carcinogenic. They have been replaced by toxaphene and lindane. Lindane was first synthesised in 1825, fifty years earlier than DDT. It has broadly similar properties to DDT, and is an effective seed-dressing for protection against attack by soil insects. It is also used as a fumigant. It is not persistent and does not create environmental problems as DDT does.

Toxaphene and lindane are used as organochlorine insecticides, having replaced similar compounds which are now banned on safety grounds.

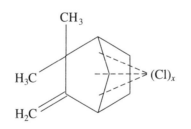

Toxaphene, a mixture of compounds produced by the chlorination of camphor

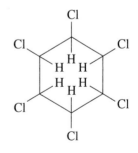

Lindane, one of the isomers of 1,2,3,4,5,6-hexachlorocyclohexane

8.4.4 ORGANOPHOSPHATE INSECTICIDES

Organophosphate insecticides undergo biodegradation ...

... e.g. parathion, which must be handled with great care as it is absorbed through the skin ...

Organophosphates have now largely replaced organochlorine pesticides. Most of the organophosphate insecticides are derivatives of phosphoric acid. They undergo biodegradation and do not accumulate so they are not a source of pollution. The first organophosphates originated from the German research work on nerve gases in the Second World War. Many of the early compounds investigated as insecticides were too toxic to mammals to be useable. A major step was the synthesis of **parathion** in 1944. Although parathion is toxic, only 120 mg being required to kill an adult human, it can be used safely if great care is taken in handling it to ensure that none comes in contact with the skin. However, several hundred people have been killed by parathion, chiefly by absorption through the skin.

*... malathion, which is
non-toxic to mammals ...
... and dichloros or
Vapona®.*

In 1950 **malathion** was synthesised and found to be non-toxic to mammals. Mammals have enzymes that can hydrolyse the ester groups and are therefore not harmed by malathion. It is widely used to control insect pests in agriculture and in the home. Dichloros, with the trade name Vapona, is more volatile and is used on strips which are hung in the house to release insecticide vapour slowly.

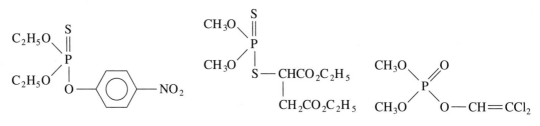

| Parathion | Malathion | Dichloros (Vapona) |

8.4.5   CARBAMATES

*Carbamates are widely
used because they are more
biodegradable than
organochlorine insecticides
and less toxic than
organophosphates.*

Carbamates are organic derivatives of carbamic acid, H_2NCO_2H. Carbamate pesticides have been used widely to replace organochlorine insecticides because some are more biodegradable than organochlorine insecticides and they are less toxic than organophosphate insecticides. They include Carbaryl, Carbofuran and Pirimicarb. The toxic effects of carbamates are due to their ability to inhibit acetylcholinesterase.

8.4.6   PYRETHRINS

*Pyrethrins are naturally
occurring compounds with
a knock-down effect on
insects. Related synthetic
compounds, e.g.
cypermethrin, are used.*

Pyrethrins are a group of naturally occurring compounds which are found in the pyrethrum flower. They have a knock-down effect on flying insects, but a second poison is needed to kill the insect. Pyrethrins have very low toxicity to warm-blooded animals. They are expensive, photosensitive and therefore unreliable. Related compounds, e.g. cypermethrin, which are not photosensitive have been synthesised and are now used on fields.

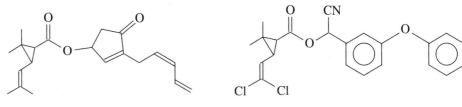

| Pyrethrin 1 | Cypermethrin |

8.4.7   ALTERNATIVES TO INSECTICIDES

*There are drawbacks to the
use of insecticides: they
have caused human deaths,
and insects develop a
resistance to them.*

Insecticides attack the nervous system, which is common to all animals. Thousands of deaths have been caused worldwide by insecticides, often by carelessness or lack of protective clothing on the part of the people using them. When plants are sprayed, only about 2% of the insecticide lands on the plants; the rest becomes an environmental pollutant. A grave drawback of insecticides is that insects become resistant to them. In the USA the use of insecticides doubled over a period of ten years without any increase in crop yields. Alternative methods of control have been developed:

Alternative methods of controlling insect populations have been investigated. They include introducing predators ...

1. Biological control: introducing insects which prey on pests. This does not give produce of the quality which the consumer demands.

2. Using viruses to kill insects. However, viruses are selective: they attack a limited number of species, and they are non-persistent and need to be introduced frequently.

3. Breeding crops that resist insects. However it sometimes turns out that the plant is more vulnerable to fungi.

... introducing viruses ...
... breeding crops that resist insects ...

4. Multiple cropping – growing a variety of crops on a piece of land – gives pest control. This works well in less intensive agriculture, but it does not fit in with high levels of agriculture and mechanical cropping. Leaving part of an area free from spraying is a compromise; it leaves a refuge for natural enemies of the pest.

... multiple cropping ...
... using pheromones to attract males and then killing them ...

5. Pheromones are used to collect males by attraction to a female odour. It is necessary to take 97% of the males out of the population to keep the population constant.

... sterilising males by radiation ...
... and using smaller applications of insecticide.

6. Sterilisation of males by radiation. When sterile males mate with females the population is wiped out. This technique only works with some insects. In most species, females can discriminate and avoid sterile males.

7. Using smaller doses. If an insecticide application does not do the job, manufacturers will compensate farmers who use their products. However, farmers must use 100% of the recommended dose to qualify for compensation. In practice, one-half to one-third of the recommended dose will work.

CHECKPOINT 8.4

1. What is meant by (*a*) a persistent pesticide, (*b*) a food chain, (*c*) a bioaccumulative pesticide?

2. (*a*) What are the advantages of organophosphates over organochlorine insecticides?

(*b*) Name two types of insecticide that are an improvement on organophosphates.

3. Briefly review methods of controlling insect pests other than commercial insecticides.

4. When insects become resistant to DDT, an enzyme converts DDT into the substance shown here.

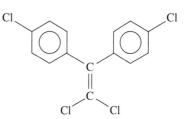

This substance is biologically inactive.

Explain how the shape of the molecule differs from that of DDT, and suggest why this might make it unable to bond to the receptor site in insects.

8.5 HERBICIDES

8.5.1 INORGANIC HERBICIDES

Until about 1960, arsenic(III) oxide and other arsenic compounds were used to kill weeds. The rates of application were high – up to hundreds of kilograms per acre. Arsenic is non-biodegradable; therefore there is the potential for arsenic pollution of surface water and groundwater from fields which have been dosed with arsenic compounds.

Arsenic compounds were widely used as herbicides.

8.5.2   PARAQUAT

Many herbicides are compounds containing two pyridine rings per molecule.

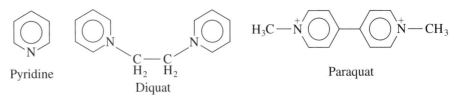

Pyridine   Diquat   Paraquat

Applied directly to the leaves, they rapidly destroy plant cells. However they bind tenaciously to soil, especially clay, so the herbicidal activity is rapidly lost and sprayed fields can be planted within a day or two of the application.

Paraquat is widely used as a herbicide although it has caused deaths through careless handling.

Paraquat is reputed to be responsible for hundreds of human deaths, through inhalation of spray, skin contact and ingestion. However, when Paraquat is used with proper procedures it is safe. It has been widely used as a herbicide, and some contamination of drinking water by Paraquat has occurred.

8.5.3   CHLOROPHENOXY HERBICIDES

The chlorophenoxy herbicides include 2,4-dichlorophenoxyethanoic acid, 2,4-D, and 2,4,5-trichlorophenoxyethanoic acid, 2,4,5-T. They were manufactured on a large scale for weed and brush control and as defoliants for military use. The intermediate for the manufacture of 2,4,5-T is 2,4,5-trichlorophenol. An accident in the manufacture of this chemical at Seveso led to one of the worst cases of land pollution by the chemical industry [see § 8.5.4].

Chlorophenoxy herbicides, e.g. 2,4-D and 2,4,5-T, are widely used.

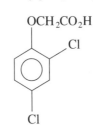

2,4-Dichlorophenoxyethanoic acid
2,4-D

2,4,5-Trichlorophenoxyethanoic acid
2,4,5-T

FIGURE 8.5A
Models of (a) 2,4-D and (b) 2,4,5-T Molecules

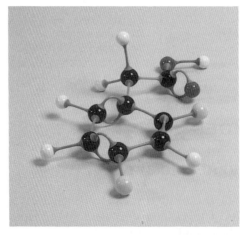

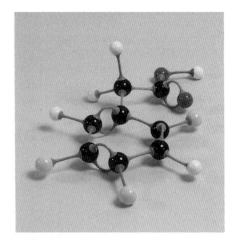

(a)   (b)

8.5.4 THE EXPLOSION AT SEVESO

The Seveso incident focused world attention on the dangers connected with the chemical industry. A chemical plant in Seveso, Italy, employed 200 people in the production of 2,4,5-trichlorophenol for the Swiss Givaudan Corporation, which is a subsidiary of Hoffman-LaRoche. On 10 July 1976 a build-up of pressure in a reaction vessel caused a safety valve to rupture, and a cloud of 2,4,5-trichlorophenol and other chemicals was released into the atmosphere.

In 1976 an explosion at a chemical plant in Seveso, Italy, released into the air a mixture of pollutants, including dioxin.

One of the chemicals released was 2,3,6,7-tetrachlorodibenzo-4-dioxin, known as TCDP and as dioxin. It is an accidental by-product of the manufacture of 2,4,5-trichlorophenol, an important intermediate in the manufacture of 2,4,5-trichloro-phenoxyethanoic acid (2,4,5-T), which is an agricultural herbicide used to control brushwood, and 'hexachlorophene', a bactericide which is used for the treatment of acne, for sterilising wounds and in skin cleansers.

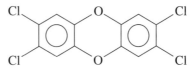

Dioxin

Dioxin is extremely toxic. Chloracne is a symptom of exposure to dioxin. Dioxin also has a teratogenic effect.

Dioxin is stable to heat, acids and alkalis, is almost insoluble in water but soluble in some organic solvents. It is 500 times more poisonous than strychnine and 10 000 times more poisonous than cyanide ion. When a person is exposed to dioxin over a long time, dioxin residues accumulate in the liver and fat cells. Symptoms are cirrhosis of the liver, damage to the heart, kidney, spleen, central nervous system, lungs and pancreas, memory and concentration disturbances and depression. The skin disease, **chloracne**, is caused by the body's attempt to get rid of the poison through the skin. Dioxin also has a **teratogenic** effect, an effect on the genes, which results in birth defects.

The population of Seveso suffered from their exposure to dioxin. Thousands of animals died. The area was so badly contaminated with dioxin that it had to be evacuated.

As a result of the accident, thousands of sheep and cows died, and people became ill, particularly with the terrible skin sores of chloracne. Nine days after the accident, people were told that the dust that had settled all over the town contained dioxin, and the town was evacuated. Among the children of Seveso, there were 134 confirmed cases of chloracne and 600 suspected cases. Of the 730 pregnant women in the town, 250 applied for abortions, and the Italian Government changed the law to allow the women to end their pregnancies.

The contaminated area of Seveso was sealed off, and experts debated how to tackle the pollution. Incineration of the contaminated soil and bacterial degradation of the dioxin in the soil were suggested. One expert recommended dismantling buildings and planting forests, rather than moving earth and washing buildings. Over a period of many years, layers of topsoil were removed and buried 10 m down beneath plastic and cement. Hoffman-LaRoche agreed to pay for all material damage and set up a fund to pay compensation to individuals.

The Seveso Directive is the EC code for controlling hazardous chemical processes.

The Seveso accident moved the European Community (EC) to bring out a set of guidelines aimed at preventing similar accidents. This legislation, known as the **Seveso Directive**, laid down regulations for the control of hazardous industrial activities in member countries of the EC.

8.6 POLYCHLORINATED BIPHENYLS, PCBS

Polychlorinated biphenyls, PCBs, were first discovered to be environmental pollutants in 1966, when they were found throughout the world in water, sediments, bird tissue and fish tissue.

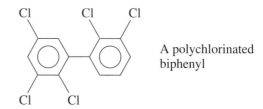

A polychlorinated biphenyl

Polychlorinated biphenyls, PCBs, have important industrial uses because they are very stable ...

... but this makes them persistent environmental pollutants, and they are no longer manufactured.

Polychlorinated biphenyls have 1 to 10 hydrogen atoms in biphenyl substituted by chlorine atoms. They have very high thermal, chemical and biological stability. They are used as fluids to provide cooling and insulation in transformers and capacitors. They are used for impregnating cotton, as plasticisers and as additives to some epoxy paints. The very fact that PCBs are so stable has led to their widespread dispersion and accumulation in the environment. In 1981, the accumulation of PCBs in wildlife led to a warning to hunters in the USA; they were warned to limit their consumption of wild ducks to two meals per month, to remove the skin and fat and discard the stuffing! The manufacture of PCBs has now been discontinued.

CHECKPOINT 8.6

1. Name three herbicides of different chemical types. Say which is the most persistent and which the least persistent.

2. The accident in Seveso started with a safety valve giving way under pressure.

(*a*) What improvements do you think chemical engineers would make to the plant after such an accident?

(*b*) Trichlorophenol vapour was released into the atmosphere. What chemical could be used to absorb trichlorophenol?

(*c*) Why do you think it took the authorities nine days to evacuate Seveso?

(*d*) Why did the Italian Government change the abortion law after Seveso?

3. How does the accident in Bhopal in 1984 [§ 3.15] compare with the accident in Seveso in 1976?

8.7 LANDFILL SITES

An increasing population, increasing affluence and an increase in packaging have led to a big increase in the amount of waste that must be disposed of.

There has been a rapid growth in the quantity of solid waste to be disposed of. One factor is the increase in the population, and a second factor is the rise in affluence: the rise in consumption per head has increased. A third factor is technology. For example, transporting beer from the producer to the consumer in a refillable beer keg does not create waste. Transporting beer in beer bottles which can be returned for refilling creates little waste. The widespread use of non-returnable bottles and cans creates tonnes of waste. The growth of self-service supermarkets has led to a great increase in packaging because an attractive package draws a customer's attention to a product. Small items are packaged to make them more difficult to shop-lift. A survey in the USA showed that 45% of paper produced, 15% of aluminium produced, 75% of glass produced and 30% of plastics went into packaging. Packaging accounts for 20–30% of household waste and 10% of commercial and industrial waste.

There are many instances where the control of air pollution or water pollution leads to new solid wastes. Fly ash is separated from the exhaust gases of coal-fired power stations. Particulate matter is separated from the effluent gases of iron and steel furnaces. These solid wastes must be disposed of. Industrial, mining and agricultural wastes are usually handled by their producers, often at the sites where they are produced. Agricultural wastes may be converted into organic matter which can be applied to soil.

Industrial waste also must be disposed of . . . . . . including waste collected by anti-pollution devices.

Municipal waste is transported to landfill sites which are subject to strict regulations. A site may be a disused quarry, a natural depression or a pit. The major concern is to prevent leachate being formed by the entry of surface water and filtering out to contaminate groundwater. A landfill is built on compacted soil of low permeability, preferably clay. The base is a flexible liner of watertight material, e.g. rubber or plastic. The liner is covered with granular material in which a drainage system is installed. On top of this is another flexible liner, and above this is a primary drainage system, a perforated pipe into which leachate can drain. The drainage system is covered with a layer of granules, and the rubbish is placed on top of this. Wastes of different kinds are separated by partitions of clay covered with liner material. When the site is full, the waste is capped to prevent surface water entering, and is covered with compacted soil.

Municipal waste is buried in landfill sites. The construction and the operation of a landfill are carefully controlled so that liquid leachate is collected and does not seep into the ground.

Organic material in a landfill is decomposed by anaerobic soil bacteria. The products include methane, water, carbon dioxide, hydrogen, hydrogen sulphide and organic acids. The organic acids can react with carbonates in surrounding soils and leach them from the soil. Methane and hydrogen usually disperse into the atmosphere without creating a hazard. In a number of cases, the combustible gases have been collected and used as fuel. Hydrogen sulphide may be formed by the reaction of sulphides with acids and by the reduction of sulphates by anaerobic bacteria. Other gases, such as carbon dioxide, are harmless, but if too much gas permeates the site it will carry with it into the atmosphere volatile aromatic compounds, and some of these compounds are toxic.

Organic matter is decomposed by anaerobic soil bacteria. Gaseous products disperse into the atmosphere.

Hazardous waste leachate can be treated by a variety of chemical methods. Excessive acidity or basicity must be neutralised before micro-organisms can thrive in the waste and digest it. Cyanide can be oxidised with chlorine. Organic material can be oxidised with ozone, hydrogen peroxide or oxygen. Heavy metals must be precipitated as oxides, hydroxides, carbonates or sulphides [see § 5.7 and § 8.14].

Hazardous leachate is treated chemically, e.g. by neutralisation, oxidation or precipitation, to render it safe.

It is becoming more and more difficult to find new landfill sites.

Most people are strongly opposed to the construction of a disposal site of any kind in their vicinity. As existing sites become filled it becomes more difficult to find acceptable locations for new landfills. Authorities are therefore turning to incineration as the means of waste disposal.

8.8 INCINERATORS

Incineration is an alternative to burial. A high temperature is needed . . .

The incineration of waste needs a temperature at which complete combustion of oxidisable material will occur. Ash, glass, metal and other materials remain. A temperature of 770–970 °C is used. In many incinerators the heat of combustion of the waste is used to help to maintain the temperature.

FIGURE 8.8A
An Incinerator

. . . but the heat of combustion can be used to maintain it.

FIGURE 8.8B
Incineration of Waste

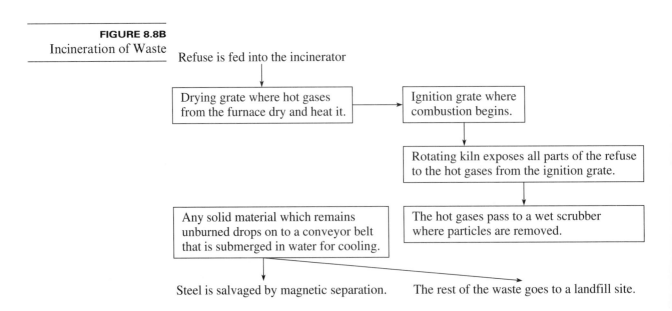

Refuse is fed into the incinerator

| Drying grate where hot gases from the furnace dry and heat it. | → | Ignition grate where combustion begins. |

Rotating kiln exposes all parts of the refuse to the hot gases from the ignition grate.

Any solid material which remains unburned drops on to a conveyor belt that is submerged in water for cooling.

The hot gases pass to a wet scrubber where particles are removed.

Steel is salvaged by magnetic separation.   The rest of the waste goes to a landfill site.

CHECKPOINT 8.8

1. Briefly describe the features of a well-constructed and carefully operated landfill.

2. Review the advantages and disadvantages of incineration compared with landfill.

3. Which of the following is least able to be disposed of by incineration? Explain your answer.

methanol, tetrachloroethene, ethanenitrile, methylbenzene, propanone

4. In some countries, waste is heated before disposal in a landfill site. Which one of the following is *not* achieved by thermal treatment of wastes?

A volume reduction

B destruction of heavy metals

C removal of volatile organic matter

D destruction of pathogens

E destruction of toxic substances

8.9 RECYCLING METALS

Recycling metals conserves metal ores and saves energy.

Metals are a valuable resource. Instead of burying metal waste it makes sense to collect it and recycle the metals. There are two savings, for instance when scrap iron is collected, melted and re-used, this saves Earth's reserves of iron ore. In addition, the energy required to mine iron ore, transport it and smelt it is several times greater than the energy required to recycle scrap iron. Aluminium is a prime candidate for recycling because, owing to its very high resistance to corrosion, used aluminium is as good as new. A second reason is the high energy consumption in the extraction of aluminium from bauxite [see *ALC*, § 19.2.2]. The cost of re-using scrap aluminium is only one twentieth of the cost of making the pure metal.

FIGURE 8.9A
Recycling Aluminium
Cans

Some firms have started to dismantle old cars, separate the metals in them, and recycle the metals.

The collection, sorting and recycling of metals has become an important industry. A major source of used metals is motor vehicles. There are about 140 million cars in Europe and 12 million a year reach the end of the road. About 75% of this mass of about 12 million tonnes is iron and steel. The European Community has encouraged member nations to make car recycling a priority. What is needed is to be able to identify components and to be able to dismantle cars rapidly. Some German car manufacturers, e.g. BMW and Mercedes, have started to design recyclability into their vehicles. They have developed **disassembly lines**. Items such as batteries, cylinder blocks, gear boxes, etc. are sent to specialist firms. A fragmenter smashes the rest of the material into small pieces. A magnet separates most of the ferrous metal. The rest passes to a second separator [see Figure 8.9B].

FIGURE 8.9B
The Eriez Cotswold
Separator

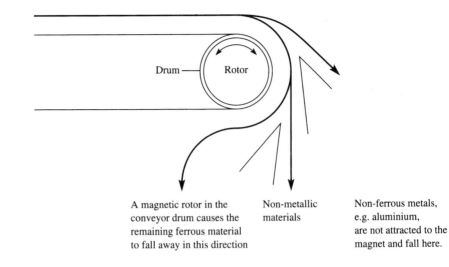

A magnetic rotor in the
conveyor drum causes the
remaining ferrous material
to fall away in this direction

Non-metallic
materials

Non-ferrous metals,
e.g. aluminium,
are not attracted to the
magnet and fall here.

The non-ferrous metals are separated by a sink-and-float process. In a tank containing a solution of high density, aluminium floats and other metals sink to be recovered later.

8.10 RECYCLING GLASS

The use of cullet in making new glass saves raw materials and also saves energy because cullet makes the mixture melt at a lower temperature. It is possible to use a larger proportion of cullet if it is available.

FIGURE 8.10A
The Chemicals Used in
Glass Making

*Waste glass can be
recycled successfully.*

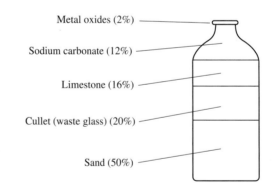

Metal oxides (2%)

Sodium carbonate (12%)

Limestone (16%)

Cullet (waste glass) (20%)

Sand (50%)

Bottle Banks collect used bottles and jars which are then sold to glass manufacturers. The manufacturers guarantee to the local authority that operates the Bottle Bank that they will buy the glass at a stated price. Benefits of the scheme are:

- saving in energy used in glass manufacture

- saving in natural resources

- saving in refuse disposal costs

- less glass litter and broken glass in the environment

*Bottle Banks collect used
glass, with savings in
natural resources and
energy and also a reduction
in domestic waste.*

- a source of income for local authorities.

8.11 DISPOSAL OF PLASTICS

8.11.1 PLASTIC WASTE

A problem with most plastics is non-biodegradability. In consequence, the disposal of plastic waste occupies more and more landfill sites.

Plastic waste constitutes about 7% of household waste. Unlike some other wastes, e.g. kitchen waste and paper, plastics are **non-biodegradable**. They are synthetic materials, and soil and water do not contain micro-organisms with the enzymes needed to feed on plastics and degrade them. During the development of plastics, the problem was different: research workers were looking for ways of making plastics last longer. They tackled the problem of photodegradation of plastics by incorporating additives in plastics to prevent oxidation and prolong the life of the plastics. Plastic waste is buried in landfill sites, and there it remains unchanged for decades. Local authorities have to find more and more landfill sites.

8.11.2 INCINERATION OF PLASTICS

Incineration of plastic waste generates useful energy.

In domestic waste, plastics help the rest of the waste materials to burn. Some plastics burn with the formation of toxic gases.

An alternative to dumping is **incineration**, with the possibility of making use of the heat generated. Plastics are petroleum products, and plastic waste contains about the same amount of energy as the oil from which it came. To burn plastic waste with the release of useful energy is an obvious solution to the problem. In the UK only about 10% of plastic waste is incinerated, but in some other countries, e.g. Denmark and Japan, incinerators consume over 70% of domestic waste. The plastic part of the waste assists in the incineration of other parts of domestic rubbish. If plastics are removed from domestic waste, together with paper, what is left is organic waste which is too wet to burn. If the waste is burnt with the plastics included, potentially useful energy is generated. Some plastics, however, burn with the formation of toxic gases, e.g. hydrogen chloride, from PVC, and hydrogen cyanide, from poly(propenenitrile), and incinerators must be designed to remove these gases from the exhaust.

8.11.3 RECYCLING PLASTICS

The EC is urging member nations to recycle used materials, and the UK Government has set a target of 50% recycling of household waste by the year 2000. Recycling is the most efficient use of resources. The municipal solid waste collected in the UK each year contains about 1.5 million tonnes of plastics. In addition, used plastics from large articles such as fridges, washing machines, agricultural machinery, cars, etc. bring the total up to about 1.8 million tonnes. A major difficulty in recycling is that, since different plastics have widely differing properties, mixed plastic waste is of limited use. While any two glass bottles can be recycled together, the same is not true of a PVC bottle and a poly(ethene) bottle. If these are melted down together, it may be possible to make some articles from the mixture, but mixtures of plastics are much weaker than individual plastics.

Recycling is an efficient method of using waste plastics, but mixed plastic waste is difficult to deal with.

The easiest plastics to recycle are off-cuts and substandard products, which consist of a known single plastic. The average car contains about 140 kg of plastics, which end up as waste when the vehicle reaches the end of its life. Refrigerators and washing machines pose a similar problem. Manufacturers have made a start in printing bar codes on plastic components to identify them and assist in sorting them into the different types of plastic waste.

Individual plastics can be recycled with ease, and some steps have been taken to collect different plastics separately.

Separating the different plastics makes recycling more worth while. The sorting of plastics into individual types of plastic is an enormous task. Some sorting is done by hand: plastic bottles can be sorted into poly(ethene), poly(chloroethene), PVC and poly(ethene benzene-1,4-dicarboxylate), PET. Mechanical separation is difficult but a process has been devised which detects the chlorine in PVC by means of X-rays, allowing PVC bottles to be separated from poly(ethene) and PET. These are then separated on the basis of their different densities. Recycled PVC is used for drains and sewer pipes, shoe soles, flooring and packaging of non-food items.

Methods of separating mixtures of plastics – manual and mechanical – have been developed . . .

In cases where collection and separation are easy, recycling is profitable. Examples are poly(ethene) from agricultural mulch films (films with holes through which the crop can grow and which must rot or be removed at the end of the growing season) and from shrink wrap packaging from supermarkets. About 10% of the poly(ethene) film we use is recycled and used to make black refuse bags etc. Telephone hand sets of poly(ethene) are collected and recycled. Poly(propene) car bumpers and casings from car batteries are recycled. Nevertheless only about 7% of the poly(propene) we use is recycled. Soft-drink bottles made of PET can be melted down and moulded into new bottles. They can also be melted down and drawn into fibres to be used in cushion filling and upholstery stuffing and carpet manufacture. All these items can be collected separately. Fast food counters use poly(styrene) containers because they offer thermal insulation. There is a UK poly(styrene) recycling organisation which targets fast food outlets.

. . . but the best results are obtained when plastics are collected separately.

With municipal solid waste, it has been estimated that no more than 50% of the plastic material can be separated, cleaned and reprocessed. Some of the mixed plastic recovered from household waste is recycled by shredding, melting and extrusion in the shape of planks. This recycled plastic can be used to construct items such as agricultural fencing, pigsties, pens and garden seats. There is a limited demand for plastics of this quality.

Recycled mixed plastic waste has limited uses.

8.11.4 PYROLYSIS

When plastics are heated in air, they burn. When they are heated in the absence of air, they are **pyrolysed** (split up by heat). The products can be separated by fractional distillation and then used in the manufacture of other materials, including plastics. The heat generated by burning a small proportion of a load of plastics waste can be used to crack the rest of the load in the furnace. Problems still to be solved include the difficulty of separating different types of plastics, and the difficulty of removing additives. The process is expensive and at present costs about five times as much as dumping the plastics in a landfill site.

Pyrolysis of plastics produces useful substances.

8.11.5 BIODEGRADABLE PLASTICS

Chemists have invented some biodegradable plastics. These may be **biopolymers**, which are made by living organisms, or **synthetic plastics**. Three types of synthetic plastics are **photodegradable plastics, synthetic biodegradable plastics** and **water-soluble plastics**.

Biodegradable plastics have been developed.

BIOPOLYMERS

In the UK, ICI markets the biopolymer poly(3-hydroxybutanoic acid), PHB, which has the trade name Biopol. It is made by certain bacteria from glucose. Biopol is used for special applications such as surgical stitches which dissolve in time inside the body. When Biopol is discarded, micro-organisms in the soil, in river water and in the body can break it down within 9 months. The properties of the polymer can be tailored by incorporating copolymers to make it suitable for a range of articles, e.g. shampoo bottles and carrier bags. At present a Biopol container is seven times the price of a poly(ethene) container. With increasing use, the price may well fall.

Biopolymers made from natural products can be broken down by micro-organisms.

PHOTODEGRADABLE PLASTICS

Photodegradable plastics break down in sunlight.

A Canadian firm has produced a photodegradable polymer, which they incorporate in polystyrene cups. Exposed to sunlight for 60 days, the cups break down into dust particles. To make plastics photodegradable, it is necessary to incorporate in them a substance that will absorb sunlight and as a result become sufficiently reactive to react with the plastic molecules. Alternatively, if a carbonyl group can be incorporated into the polymer chain, the carbonyl group will absorb light and use the energy to break chemical bonds in the polymer. Introduction of carbonyl groups is done by polymerising the monomer, e.g. ethene, with carbon monoxide. Polymers containing 1% of carbonyl groups lose their strength after 2 days of sunlight, compared with about 300 days in the absence of carbonyl groups.

SYNTHETIC BIODEGRADABLE PLASTICS

Synthetic biodegradable plastics incorporate natural materials, e.g. starch, in the plastic so that the natural material will be digested by micro-organisms.

An Italian company, Feruzzi, has produced a biodegradable polymer which is suitable for carrier bags. The material consists of poly(ethene) and up to 50% starch. The poly(ethene) chains and starch chains interweave to form a material which is strong enough for shopping bags. When the material is buried, micro-organisms begin to feed on the starch, converting it into carbon dioxide and water, and in time the polymer chains dissolve in water. The cost at present is about twice that of a regular plastic bag. Europe uses 100 000 tonnes of degradable plastic a year, and the consumption is rising.

SOLUBLE PLASTICS

Soluble plastics can be designed to dissolve slowly in hot water, warm water or cold water.

Plastics which dissolve in water can be designed to be soluble in cold water, in warm water or only in hot water. Poly(ethenol), $(\text{—CH}_2\text{—CHOH—})_n$, also called poly(vinyl alcohol), is an example. It is used as a packaging for swimming pool chemicals, descalers, seed strips and other uses to a total of 100 tonnes a year in the UK.

8.11.6 RECYCLING VERSUS BIODEGRADABILITY

The governments of USA, Sweden and Italy have passed laws making degradability compulsory for plastics in certain types of packaging. On the other hand, Friends of the Earth and British Plastics Federation now oppose degradable plastics on the grounds that it is better to recycle plastic waste.

Some plastic waste is never collected, e.g. plastics used for agricultural purposes and plastics thrown overboard from ships. There is much to be gained from using biodegradable plastics for such articles. Other plastic objects could be collected for recycling. Some plastics are easy to recycle, e.g. poly(ethene benzene-1,4-dicarboxylate), PET, from which many bottles and jars are made There are problems in recycling plastics, however, as described in § 8.11.3.

There is debate on the merits of biodegradable plastics versus recyclable plastics.

There are two solutions to the problem of plastic waste. One is to make degradable plastics, and the other is to recycle plastics. The two solutions do not live well together. Although some biodegradable plastics, e.g. PHB, can be recycled with other plastics, photodegradable plastics cannot be included. Waste plastics can be turned into items such as sacks, park benches, roofing and drain pipes. You can imagine the accidents that could occur if such materials were to break up in sunlight.

ICI has made a new grade of PET called 'Melinar' that is used to make plastic bottles for soft drinks. The bottles can be used up to 20 times over a period of 5 years before they are recycled into fillings for toys, anoraks and duvets.

FIGURE 8.11A
Melinar Bottles

8.12 RECYCLING PAPER

Pulping and recycling used paper is a simple matter. The recycled paper is softer and less strong. Much more used paper could usefully be recycled.

High quality paper used for books is made by chemical pulping, in which lignin is dissolved from the cellulose. Low-quality paper, used for newspapers, is made by mechanically grinding wood in water. There is no way of separating high-quality and low-quality paper during recycling so the recycled paper is rather off-colour and weaker than the original paper. Many synthetic materials are incorporated in paper: coatings of latex and clay, resins, wax, glues, dyes and pigments. When paper disintegrates in water for repulping, hairlike fibrils attached to the cellulose fibres are damaged. This gives the paper a softer feel and reduced strength. The quantity of waste paper we amass would supply 60% of our paper consumption, but only one third of it is recycled. However about 25% of the paper produced comes from waste wood reclaimed from sawmills etc.

═══════════════════════════ **CHECKPOINT 8.12** ═══════════════════════════

1. Millions of plastic bags are discarded after one or two hours' use. Many plastic bags are made of poly(ethene).

(*a*) Explain how poly(ethene) is obtained from petroleum.

(*b*) How long did petroleum take to form?

(*c*) Can it be replaced?

(*d*) What is meant by the statement that plastic bags are non-biodegradable? What significance does this statement have for the disposal of plastic waste?

2. Gas is used to convert polystyrene into polystyrene foam.

(*a*) What is the advantage of polystyrene foam for serving food in take-away restaurants?

(*b*) For how long is a polystyrene foam package in use?

(*c*) What happens to the polystyrene it contains?

(*d*) What happens to the gas it contains?

3. (*a*) Explain the advantage of using a dissolving polymer for (i) surgical stitches, (ii) a laundry bag used to store laundry in a hospital where there is danger of infection.

(*b*) Give three uses for which a dissolving plastic would be unsuitable.

4. Suggest (*a*) applications for a photodegradable plastic,

(*b*) items for which it would not be suitable.

5. Discuss the pros and cons of recycling plastics or using biodegradable plastics.

6. (*a*) Mention four areas in which savings are made when glass is recycled.

(*b*) Why is glass easier to recycle than plastics?

7. (*a*) Mention reasons for recycling metals.

(*b*) Why is aluminium especially suitable for recycling?

(*c*) How can aluminium be separated from (i) iron, (ii) zinc?

8.13 ACCIDENTAL POLLUTION

From time to time there is an accident in the chemical industry which causes serious pollution of air and land.

The industrial accident at Seveso which caused severe and long-lived pollution of land and air was described in § 8.5.4. The industrial disaster at Bhopal which released poisonous gases into the atmosphere was described in § 3.15. These incidents of pollution by chemical industry are rare.

8.13.1 TRANSPORTING CHEMICALS

The transporting of chemicals by road, rail and sea is a potential source of pollution. The chemical industry manages the job efficiently ...

The chemical industry takes great care to avoid accidents, both in the plant and in transporting the goods to the consumer. The difficulties that must be solved in transporting corrosive chemicals safely are illustrated by bromine. Bromine is obtained from sea water and has to be transported from the plants in which it is produced, e.g. in Anglesey, to the factories where it is used, e.g. to make halobutyl rubber. The transport of bromine by road, rail or sea requires great care. Most of it is carried in lead-lined steel tanks which hold several tonnes of bromine. International regulations control the design and construction of road and rail tankers, and the safety record is good.

... as is seen from the precautions taken with bromine.

Bromine is also shipped from the Dead Sea in Israel to the petrochemical plant at Fawley near Southampton in the UK. At Fawley the tanks of bromine must be emptied. Nitrogen is led into the tank to pump bromine through plastic pipes into a storage cylinder of glass-lined steel. The cylinder is supported above a water-filled tank which catches any spillage of liquid bromine. From the storage tank, bromine is pumped through plastic pipes to the plant. A bromine detector in the area monitors the bromine concentration in the air to ensure that it does not rise above 0.1 ppm. Operators must wear protective clothing and breathing apparatus.

THE UK CHEMICAL INDUSTRY'S BLACKEST DAY

The UK disaster 20 years ago in Flixborough occurred when cyclohexane leaked from a cracked pipe. It ignited and caused an explosion which killed and injured people and damaged buildings.

Nylon-6 is made from caprolactam, $HN-(CH_2)_5-C=O$ with an O bridging. The manufacture of caproplactam was responsible for the biggest disaster in the history of the UK chemical industry. It happened in 1974 at the Flixborough works of Nypro (UK) in Lincolnshire. Part of the process involves the oxidation of cyclohexane to cyclohexanol and cyclohexanone. One of the reactors in the plant had been removed for repair and replaced temporarily by a pipe of large diameter. A crack formed in the pipe and cyclohexane leaked out to form a cloud of highly flammable vapour. This ignited. The resulting explosion killed 28 people and injured 100 others as well as damaging 2000 houses and shops. Had the explosion not occurred on a Saturday, when few people were working in the plant, the loss of life would have been much greater.

At the time of the accident there was no plant engineer because the post was vacant.

8.14 TREATMENT OF INDUSTRIAL WASTE

8.14.1 WASTEWATER

The treatment of industrial wastewater has been covered in § 5.7.

8.14.2 LEACHING

A hazardous component can be removed from solid waste by means of a chemical reaction which converts it into a soluble compound.

1. **Acids** will dissolve a number of metal salts of low solubility, e.g.

$$PbCO_3(s) + H^+(aq) \longrightarrow Pb^{2+}(aq) + HCO_3^-(aq)$$

Acids also dissolve basic organic compounds such as amines and phenylamine. Weak acids such as ethanoic acid are the safest to use. Care must be taken over extraction with acids because if cyanides and sulphides are present toxic hydrogen cyanide and hydrogen sulphide are formed.

2. **Chelating agents**, e.g. EDTA (represented as HY^{3-}) dissolve insoluble metal salts by forming soluble complexes.

$$FeS(s) + HY^{3-}(aq) \longrightarrow FeY^{2-}(aq) + HS^-(aq)$$

3. **Reducing agents**: Heavy metal ions in soil contaminated by hazardous wastes may be associated with insoluble compounds such as iron(III) oxide and manganese(IV) oxide. The action of reducing agents is to reduce these oxides to soluble salts of iron(II) and manganese(II). At the same time the heavy metal ions may dissolve and be removed with the aqueous solution.

4. **Photolysis** can be used to destroy a number of kinds of hazardous wastes. Hazardous wastes that can be destroyed by photolysis with UV light include dioxin [see §8.5.4], 2,4,6-trinitromethylbenzene, TNT, chlorinated biphenyls, PCBs, and some herbicides.

8.15 DUMPING AT SEA

Instead of polluting the land, sometimes industrial waste, sewage and sewage sludge are dumped at sea. There is a limit to the quantity of such material that the oceans can cope with, especially since dumping is often done in estuaries or fairly close to the coast.

In coastal areas, sewage is discharged into the sea after primary treatment. If it is sufficiently dilute it is broken down by aerobic bacterial action without any serious pollution. Nitrogen compounds and phosphorus compounds are oxidised to nitrates and phosphates. As populations increase the discharge of sewage into the sea becomes less safe. In 1988 off the west German coast a 3 m thick blanket of algal bloom covered the beaches of the resort island of Sylt.

The UK has until now dumped sewage sludge into the North Sea and the Irish Sea, but EC rules will phase out sea dumping by 1998. In Hong Kong some digested sewage sludge is dumped at sea and some is disposed of in landfill sites [see §5.6].

The chimneys of coal-fired power stations remove particles from the exhaust gases to avoid pollution from smoke. The solid waste that is removed from the chimneys is called fly ash. It contains oxides of silicon, iron, aluminium and other metals. It is often dumped at sea, where it sinks to form a layer on the sea bed. This can smother plants and animals on the sea bed.

8.16 DISPOSAL OF RADIOACTIVE WASTE

There are various types of radioactive waste. Some are the radioisotopes which scientists use for research in laboratories and for treating patients in hospitals. Some are the used fuel rods from nuclear power stations. The methods of disposing of the waste depends on whether it is of low-level, high-level or intermediate-level activity and on how long the waste will remain radioactive. The time taken for the radiactivity of a radioisotope to decay to half its initial value is the **half-life** of the isotope. The half-lives of the radioisotopes to be stored vary enormously; for example:

iodine-131, 8 days

strontium-90, 28 years

caesium-137, 30 years

plutonium-239, 2.4×10^4 years

caesium-135, 3×10^6 years

uranium-235, 7×10^8 years

Radioactive waste poses a special problem. Some radioisotopes have very long half-lives, e.g. 240 000 years for plutonium-239.

Plutonium-239 is one of the major products of reprocessing the fuel from nuclear power stations. It must be stored for 24 000 years before its radioactivity has fallen by a half. After a further 24 000 years, the radioactivity will have fallen to a quarter of the initial value. After 100 000 years, the radioactivity will still be one-sixteenth of its initial value. The measures which we take for storing long-lived radioactive waste will be handed to future generations for them to maintain and safeguard.

8.16.1 LOW-LEVEL RADIOACTIVE WASTE

Low-level waste includes items such as paper towels, rubber gloves and protective clothing that have been used in areas where radioactive materials are handled. Skips of low-level radioactive waste are sent by road and rail from all over Great Britain to the disposal site at Drigg in Cumbria, which is owned and operated by British Nuclear Fuels plc, BNFL. Originally the waste was placed in clay-lined trenches. Since 1989 the waste has been placed in steel containers inside concrete-lined vaults. Since 1959, 750 000 m³ of waste has been safely disposed of at Drigg. When the site is filled, sometime in the twenty-first century, it will be covered with a waterproof membrane and then landscaped.

Low-level radioactive waste is disposed of inside steel containers in a concrete-lined vault in Cumbria.

FIGURE 8.16A
Low-level Radioactive Waste Repository at Drigg

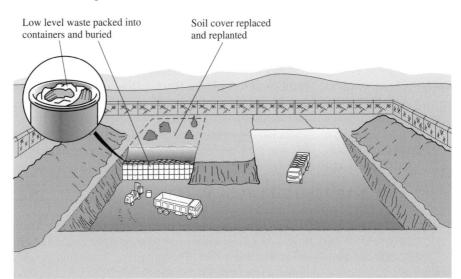

Low level waste packed into containers and buried

Soil cover replaced and replanted

Spent nuclear fuel rods are hot because radioactive decay generates heat. They are stored and cooled in vast tanks of water called 'ponds'. The cooling pond water becomes slightly radioactive. It was discharged into the Irish Sea for many years, but there was so much public concern over this practice that BNFL constructed a plant to treat the water. This is the Site Ion Exchange Effluent Plant which filters the water through an ion exchange resin that removes strontium and caesium, which are responsible for much of the radioactivity. Then the water is carefully monitored before being pumped into the sea.

Radioactive strontium and caesium are removed from the 'pond water' used to cool fuel rods before it is discharged into the sea.

8.16.2 INTERMEDIATE WASTE

Radioactive waste of intermediate activity, including the metal cladding from spent fuel rods, is packed into steel drums which are placed in concrete frames. Intermediate waste of short half-life is buried in a deep pit.

Nuclear fuel rods are encased in magnesium alloy cans. When spent fuel is taken from the reactor for reprocessing, the metal is stripped off the rods. The metal is one type of intermediate-level radioactive waste. Other types are contaminated equipment and sludges from various treatment processes. The UK used to dispose of this type of waste at sea but this practice has been stopped. It is now set in concrete and packed into steel drums. The drums are placed in concrete frames. About 3000 m^3 of waste has been packaged in this way and is stored at Sellafield and other sites. It can be kept safely in this kind of interim storage for decades. To keep it safe for thousands of years the waste containers will be placed in rock caverns deep underground. A repository for intermediate waste of short half-life is shown in Figure 8.16B.

FIGURE 8.16B
Repository for Short-lived Intermediate Waste

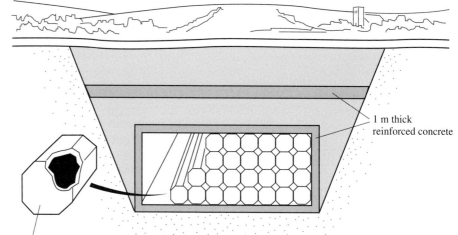

1 m thick reinforced concrete

Waste filled drums cast in concrete blocks

Intermediate waste of long half-life is being stored at Sellafield until a permanent repository in a deep mine has been constructed to hold it.

For intermediate waste of longer-lived isotopes, a deep mine is planned, hundreds of metres underground. Rocks will provide an additional barrier between the waste and the environment. The repository will accommodate all the intermediate waste produced by the nuclear industry for many decades. Nirex, the company which is responsible for the disposal of radioactive waste in the UK, has applied for planning permission to build the Rock Lab near Sellafield to study the rock formation and decide whether a safe repository could be built in the area – a repository that would be safe for thousands of years. Permission has been refused by Cumbria County Council. Nirex has appealed to the Secretary of State for the Environment against the decision because the plan is to have a repository in operation by 2010.

8.16.3 HIGH-LEVEL WASTE

Highly radioactive waste comes from the reprocessing of spent nuclear fuel. It is vitrified.

High-level waste arises from the reprocessing of spent nuclear fuel. During reprocessing 96% of the uranium is recovered for re-use. About 1% is plutonium, which is a valuable nuclear fuel. The remaining 3% is high-level liquid waste. The highly active liquid waste is vitrified in the Sellafield vitrification plant.

FIGURE 8.16C
Repository for Long-lived
Intermediate Radioactive
Waste

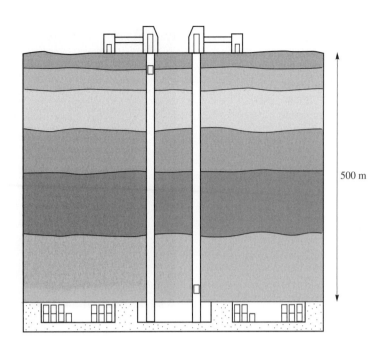

500 m

*The vitrified waste is
constantly cooled to
remove the heat generated
by its radioactivity.*

1. The liquid waste is sent through a rotating tube inside a furnace and emerges as a
dry powder.

2. The powder is fed into a melting pot together with glass-making material.

3. The molten material is poured into containers.

4. The containers are transferred to an air-cooled store and stacked in stainless steel
tubes. Air flows round the containers constantly to remove the heat generated by the
radioactivity of the vitrified material.

*A permanent repository
has to be constructed to
contain the vitrified waste.*

5. After about 50 years the vitrified waste will have cooled sufficiently to be put into a
permanent repository.

8.16.4 TRANSPORT OF RADIOACTIVE WASTE

Spent nuclear fuel is transported to Sellafield in Cumbria for reprocessing. This
practice has caused some concern to the public who could envisage the results of an
accident in which the container was fractured and radioactive material released.
Radioactive material is transported in 'flasks'. A 'flask' is made of steel and weighs up
to 110 tonnes.

FIGURE 8.16D
Spent Fuel Transport
Flask

*Spent nuclear fuel is
transported in huge steel
'flasks'.*

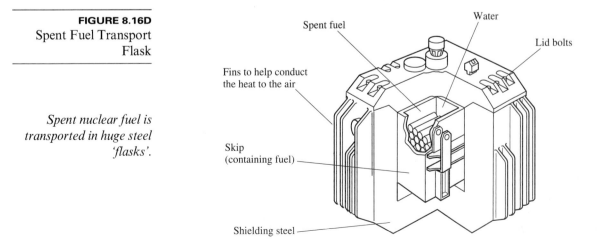

Spent fuel

Water

Lid bolts

Fins to help conduct
the heat to the air

Skip
(containing fuel)

Shielding steel

FIGURE 8.16E
Fuel Transport Flasks
Arriving at the Sellafield
Fuel Handling Plant

*In a test crash, one of these
flasks withstood the impact
of a heavy train travelling
at 160 kmh⁻¹.*

In 1984 a worldwide TV audience of millions watched a diesel train on its last run. At 140 tonnes, one of the heaviest used by British Rail, it pulled three coaches of 35 tonnes each. The train was travelling at 160 km h^{-1} when it hit a nuclear fuel flask [see Figure 8.16F]. The crash was spectacular with an explosion of fire, and it took 5 seconds for the train to come to rest. The engine was crushed, the carriages were a write-off, and the flask was completely intact. The Central Electricity Generating Board had made its point; the flasks are safe.

FIGURE 8.17F
The Crash

CHECKPOINT 8.16

1. (*a*) What sort of articles make up low-level radioactive waste? What happens to radioactive waste of this kind?

(*b*) What radioactive waste of intermediate level comes from nuclear power stations? What happens to this waste?

(*c*) What is the source of high-level radioactive waste from nuclear power stations? What happens to this waste?

2. Explain the difficulty in disposing of radioactive waste that does not apply to other kinds of waste.

3. Why are people worried about the transport of radioactive waste? What has been done to reassure the public over this?

QUESTIONS ON CHAPTER 8

1. A sanitary landfill is a means of disposing of waste that has less impact on the environment than dumping on land or in the ocean. Describe how such a landfill should be managed.

2. Municipal waste is disposed of in incinerators. Discuss the advantages and disadvantages of this method compared with landfill sites.

3. (*a*) Describe two kinds of soil pollution.

(*b*) Explain why organic pollutants tend to remain in the top layer of soil while inorganic pollutants are more rapidly leached into waterways.

4. What does PCB stand for? Give the formula of one PCB. For what purpose were PCBs used? Why has their use been discontinued?

5. The following are measures for dealing with wastes:

(*a*) reducing the volume of waste by incineration

(*b*) placing waste in landfills protected from leaching

(*c*) treating waste to render it innocuous

(*d*) reduction of wastes at the source

(*e*) recycling as much waste as practicable.

Place the measures in decreasing order of importance and discuss your reasons for doing so.

6. What are the advantages of disposing of hazardous wastes above ground, as opposed to burial in a landfill?

7. What are the special problems in disposing of radioactive waste?

ANSWERS

CHAPTER 1: THE ENVIRONMENT

Questions on Chapter 1
1. See § 1.1.1.
2. See § 1.3.
3. See § 1.5.
4. See § 1.8.
5. See § 1.9.

CHAPTER 2: THE ATMOSPHERE

Checkpoint 2.3
1. The force of gravity holds the atmosphere close to the Earth's surface. With increasing altitude, the force of gravity decreases and the concentration of gases decreases.
2. As the pressure outside the balloon decreases as the balloon ascends, the volume of gas inside it increases.
3. (a) The concentrations of H_2O, CO_2, etc. which absorb IR radiation from Earth's surface decrease with height.
   (b) Absorption of UV radiation by ozone
   (c) The concentrations of species that can absorb radiation are low.
   (d) Ions and radicals are formed by the absorption of radiation.

Checkpoint 2.4
1. See §§ 1.8 and 2.4.
2. Water is always present. Sunlight photolyses O_2 to O^* which reacts with H_2O to form ·OH; see § 2.4.
3. The NO_2 molecule has an unpaired electron [see § 2.4] which makes it combine rapidly with free radicals. It is able to absorb light over a range of wavelengths to form an electronically excited molecule.

Checkpoint 2.8
1. See § 2.6 and Figure 2.6A.
   (a) The Haber process, vehicle engines
   (b) Harvesting crops, not returning excreta to the soil
   (c) Applying fertilisers to the soil
2. See § 2.7.
3. See § 2.7.
4. $CaCO_3$ could dissolve to put $CO_3{}^{2-}$ ions into the ocean, $CO_3{}^{2-}$ could form $CO_2(g)$ which takes part in photosynthesis with the formation of oxygen.
5. For nitrifying and denitrifying bacteria, see §§ 2.6.1, 2.6.2 and Figure 2.6A.
6. (a) Carbon dioxide is used in photosynthesis and formed in respiration, while oxygen is formed in photosynthesis and used in respiration.

(b) Plants photosynthesise more in daylight than at night-time and more in spring than in autumn.

Checkpoint 2.9
1. See § 2.9 for an explanation of how the greenhouse effect makes the Earth warm enough to support life.
2. (a) The exposed land would radiate energy back towards space.
   (b) Water vapour is a greenhouse gas.
   (c) Carbon dioxide would come out of solution in the oceans.
   (d) Carbon dioxide and water vapour from volcanic action would add to the greenhouse cover.
3. (a) Cutting down trees leaves fewer to remove carbon dioxide from the atmosphere in photosynthesis.
   (c) Some financial incentive, e.g. payment from richer countries or cancellation of debts, in return for a stop to deforestation.
4. (a) See § 2.9.1.
   (b) See § 2.9.
   (c) Accurate measurements can be made at present, but there is doubt over the accuracy of measurements of past temperatures; see § 2.9.2.
   (d) (i) $6CO_2 + 6H_2O \longrightarrow C_6H_{12}O_6 + 6O_2$ and
       (ii) $CO_2(g) + aq \longrightarrow CO_2(aq)$;
           $CO_2(aq) + CaO(s) \longrightarrow CaCO_3(s)$
5. (a) More photosynthesis takes place in summer.
   (b) At the Equator there is less difference between summer and winter. At the Poles there is little vegetation.
   (c) The increase in the combustion of fossil fuels and the decrease in the area of forest.
   (d) See § 2.9.5 for the threat of global warming.

Checkpoint 2.10
1. (a) It absorbs dangerous UV radiation; see § 2.10.
   (b) A supply of O_2 and UV light is needed to make O_3. Reactions with naturally occurring radicals, e.g. H·, NO·, O·, remove O_3.
2. See equations in § 2.10.
3. (a) N_2O (b) HNO_3

Questions on Chapter 2

1. (a) The temperature falls with altitude in the troposphere because the concentration of species that can absorb energy decreases. Then the temperature increases in the stratosphere as the intensity of UV light leads to the formation of $O \cdot$ radicals and the exothermic reaction $O \cdot + O_2 \longrightarrow O_3$ [see § 2.10].
   (b) The mesosphere lacks sufficient species to absorb radiation and the temperature falls.
2. See nitrification § 2.6.1 and denitrification § 2.6.2 and Figure 2.6A.
3. (a) See § 2.8 and Figure 2.8A.
   (b) Photosynthesis and respiration are inter-related; see Figure 2.7A.

(c) Anaerobic respiration; see § 2.7.
4. (a) A third body absorbs energy in an exothermic reaction and prevents the product from dissociating; see § 2.4.
   (b) an unpaired electron and an excited species; see § 2.4.
   (c) NO_2^* could lose energy by collisions; the other species need to react to share their unpaired electrons.
5. UV light is intense. There is a high concentration of free radicals. Photochemical reactions between free radicals and contaminants occur.
6. $\cdot O$ and O_3
7. Anaerobic respiration: $2(CH_2O) \longrightarrow CH_4 + CO_2$
   Aerobic respiration: $(CH_2O) + O_2 \longrightarrow CO_2 + H_2O$
   $15.0\,dm^3$ CH_4 therefore $30.0\,dm^3$ O_2

CHAPTER 3: AIR POLLUTION

Checkpoint 3.1
1. For a certain mass of particles, the ratio surface area/volume is larger for smaller particles.
2. Smaller particles can penetrate to the lungs.
3. See § 3.1.
4. See § 3.1.2.

Checkpoint 3.3
1. (a) Carbon monoxide combines with haemoglobin. See §3.3 for its effects. Being colourless and odourless, it gives no warning of its presence.
   (b) $\cdot OH$ radicals and soil micro-organisms; see § 3.3.
   (c) The internal combustion engine
2. (a) Prevents knocking.
   (b) Atmospheric pollution; see § 3.3.
   (c) Cars were made to run on unleaded petrol.

Checkpoint 3.6
1. (a) Cause damage to the respiratory system and increase asthma attacks. Contribute to the formation of acid rain. Contribute to photochemical smog.
   (b) Dissolve in water droplets in the atmosphere and precipitate in the rain.
2. See § 3.5.2.
3. NO_2 has an unpaired electron. See § 3.5.1.
4. NO_x are converted into nitric acid; see § 3.5.3.
5. If 57% of the waste sulphur from this power station could be saved, it would equal the annual tonnage of sulphur imported.
6. (a) The FGD method uses less limestone to produce a certain mass of $CaSO_4$.
   (b) The same
   (c) The PFBC process prduces more waste to be disposed of.

Checkpoint 3.8
1. **B**
2. Oxygen atoms, $O \cdot$, ozone molecules, O_3
3. See § 3.8.1.
4. See § 3.8.2.
5. Step 3 in the formation of photochemical smog requires hydrocarbons; see § 3.8.

6. Sulphur dioxide
7. **D**
8. (a) sunny, windless, cloudless
   (b) valleys
   (c) Refer to Figure 3.8A. Stagnant warm air and sunlight lead to photochemical smog.

Checkpoint 3.9
1. (a) (i) In acceleration, the pistons move up and down in the cylinders more rapidly and there is less time for complete combustion.
   (ii) During idling, the air flow is insufficient for complete combustion.
   (b) The air flow decreases and combustion is incomplete.
   (c) The rate at which the mixture in the cylinders fires increases during acceleration and oxidation of nitrogen to NO_x increases.
2. Pollution of the air is reduced, and pollution of land by precipitation of lead compounds from the air is reduced. Catalytic converters can be fitted to cars.
3. 10 kg TEL. The lead was emitted as bromides of lead with exhaust gases.
4. An increased demand for electricity means that more coal would be burnt in power stations, with increasing atmospheric pollution. However, the emission from a power station is easier to control than that from thousands of individual cars.

Checkpoint 3.10
1. (a) Formation of NO_x.
   (b) See § 3.9.3.
   (c) See § 3.9.2.
2. See § 3.10.1. If the country has no oil, there is special interest. A country must have land available to grow sugar cane and sunlight to ripen it.
3. See § 3.10.2.
4. See § 3.10.4. (a) The combustion products are harmless.
   (b) Pure hydrogen burns smoothly, but if there is air in the stream of hydrogen it burns explosively. Great care must be taken over using it.

Checkpoint 3.11

1. (*a*), (*b*), (*c*) See § 3.11.3.
   (*d*) See § 3.11.4.
2. (*a*) See § 3.11.1, equations (*a*) and (*c*).
   (*b*) See § 3.11.1.
   (*c*) See § 3.11.2.
   (*d*) They are very unreactive because C–Halogen bonds are very strong.
3. (*a*) See § 3.11.2.
   (*b*) See § 3.11.5.
4. (*a*) CFCs will remain in the atmosphere for a very long time; see § 3.11.2.
   (*b*) The cost of switching to dearer alternatives makes the poorer nations reluctant to change.
   (*c*) Some financial incentive

Checkpoint 3.12

1. (*a*) Vehicle traffic in the rush hour.
   (*b*) NO is subsequently oxidised to NO_2.
   (*c*) The evening rush hour generates NO.
   (*d*) Sunlight is needed for the photolytic reactions which produce ozone.

Checkpoint 3.14

1. (*a*) There is more circulation of air in daytime, with windows and doors being opened frequently.
   (*b*) In winter windows are closed and draughts are minimised; therefore radon is less likely to escape than in the summer.

Questions on Chapter 3

1. (*a*) See § 2.9.
   (*b*) Particles reflect the Sun's radiation and exert a cooling effect on the Earth.
   (*c*) The level will rise; see § 2.9.5.
2. (*a*) They contribute to the formation of photochemical smog; see §§ 3.7.1, 3.8.
   (*b*) See § 3.1.1.
3. (*a*) (i) With the engine idling the supply of air is insufficient for complete combustion, and CO rather than CO_2 is formed.

   (ii) With the engine idling the supply of air is insufficient for complete combustion, and hydrocarbons are emitted.
   (iii) With the engine accelerating, the air intake increases and the temperature of the cylinder increases and NO_x are formed.
   (*b*) NO_2
   (*c*) See § 3.8.
   (*d*) See § 3.9.2.
4. (*a*) See § 3.11.1.
   (*b*) (i) See § 3.11.1.
   (ii) See § 2.9.
5. (*a*) See § 3.8 and Figure 3.8A.
   (*b*) Los Angeles is in a valley and receives plenty of sunshine.
   (*c*) Air does not rise by convection. Pollutants accumulate at ground level.
   (*d*) Time is needed for NO to be oxidised to NO_2 and sunlight is necessary for the formation of photochemical smog; see § 3.8.
6. (*a*), and (*b*) See § 2.10.
   (*c*) See § 3.11.2.
   (*d*) See §§ 3.11.1, 3.11.2.
   (*e*) See § 3.11.3.
7. (*a*) ozone
   (*b*) chain-termination; see § 3.8, Step 5
   (*c*) They contribute to photochemical smog; see § 3.8, Step 3.
8. (*a*) N (*b*) SO_2 (*c*) oxidants
9. (*a*) It is formed by the reactions in the atmosphere of primary pollutants, e.g. sulphur dioxide and oxides of nitrogen.
   (*b*) Oxides of nitrogen are emitted in vehicle exhausts. Oil refineries emit sulphur dioxide. These gases combine with water in the atmosphere to form acid rain.
10. (*a*) Clouds form, and reaction at the surface of water droplets increases the rate of formation of $ClO\cdot$ from 'chlorine reservoirs' of $ClNO_3\cdot$ see § 3.11.3.
   (*b*) See § 3.11.4.

CHAPTER 4: THE HYDROSPHERE

Checkpoint 4.1

1. Evaporation over the ocean and condensation over the land; see Figure 4.1A.
2. Water takes in heat to vaporise and water vapour releases heat when it condenses; see Figure 2.2A.
3. Carbon dioxide dissolves in rainwater, making it weakly acidic. Sulphur dioxide and nitrogen dioxide dissolve in rainwater to produce the strong acids, sulphuric acid and nitric acid; see § 3.6.

Checkpoint 4.7

1. See § 4.3 .
2. (*a*) Oil is a mixture of hydrocarbons which cannot form hydrogen bonds with water; therefore hydrocarbon molecules cannot displace water molecules in the hydrogen-bonded association of water molecules that constitutes liquid water.

   (*b*) Ethanol molecules can form hydrogen bonds with water molecules; see Figure 4.4A.
   (*c*) Water molecules bond to ions and energy is released – the energy of hydration. This compensates for the energy required to break up the structure of the ionic compound.
   (*d*) Organic compounds dissolve in water if they can form hydrogen bonds to water; see Figure 4.4A.
3. Energy is released when bonds form between water molecules and cations and between water molecules and anions; see § 4.5.
4. (*a*) $K_{sp} = [Ca^{2+}(aq)] [SO_4^{2-}(aq)]$
   (*b*) $K_{sp} = [Ag^{+}(aq)]^2 [SO_4^{2-}(aq)]$
   (*c*) $K_{sp} = [Fe^{3+}(aq)] [OH^{-}(aq)]^3$.
5. (*a*) 7.1×10^{-5} mol dm^{-3}
   (*b*) 7.1×10^{-3} g dm^{-3}
6. 6.3×10^{-14} mol^2 dm^{-6}
7. (*a*) Soluble $[Cu(NH_3)_4]^{2+}$ ions are formed.
   (*b*) Soluble $[CuCl_4]^{2-}$ ions are formed.
   (*c*) Soluble $[Cu(CN)_4]^{2-}$ ions are formed.

Checkpoint 4.9

1. (a), (b), (c) See § 4.8.
2. The extent of hydrogen bonding between molecules is higher in water than in ethanol.
3. Loss of heat in (i) transpiration   (ii) perspiration
4. See § 4.9.
   Volume $= 2000 \times 1000 \times 100 \, m^3 = 2.0 \times 10^8 \, m^3$
   Mass $= 2.0 \times 10^8 \, g$, $\Delta T = 1 \, °C$
   Heat $= 4.2 \, J \, K^{-1} \, g^{-1} \times 2.0 \times 10^8 \, g \times 1 \, K$
   $= 8.4 \times 10^8 \, J = 840 \, MJ$

Checkpoint 4.12

1. (a) $pH < 7$ owing to $NH_4^+ + H_2O \rightleftharpoons NH_3 + H_3O^+$
   (b) $pH > 7$ owing to $CN^- + H_2O \rightleftharpoons HCN + OH^-$
   (c) $pH < 7$ owing to
      $CH_3CO_2^- + H_2O \rightleftharpoons CH_3CO_2H + OH^-$
2. (a) Hydrolysis produces H_2S by
      $S^{2-} + 2H_2O \rightleftharpoons H_2S + 2OH^-$
   (b) Hydrolysis produces HCN by
      $CN^- + H_2O \rightleftharpoons HCN + OH^-$
   (c) $2H^+(aq) + CaCO_3(s) \longrightarrow Ca^{2+}(aq) + CO_2(g)$
      $+ H_2O(l)$

Checkpoint 4.14

1. (a), (b), (c) See § 4.14.
2. (a) Calcium carbonate or magnesium carbonate; see equation in § 4.13.1.
   (b) See §§ 4.13.1, 4.13.2.

Checkpoint 4.15

1. See §§ 4.15.3, 4.15.4.
2. (a) $CO_2(aq)$, $HCO_3^-(aq)$, $CO_3^{2-}(aq)$
   (b) $HCO_3^-(aq) + H^+(aq) \rightleftharpoons CO_2(aq) + H_2O(l)$
      $CO_3^{2-}(aq) + H^+(aq) \rightleftharpoons HCO_3^-(aq)$
      $OH^-(aq) + HCO_3^-(aq) \rightleftharpoons H_2O(l) + CO_3^{2-}(aq)$
   (c) (i) low pH   (ii) neutral pH   (iii) high pH
   (d) $HCO_3^-(aq)$
   (e) $CaCO_3(s)$ dissolves to form $HCO_3^-(aq)$, $[H^+]$ falls as $H^+(aq)$ reacts with $HCO_3^-(aq)$ to form $CO_2(aq)$. $CO_2(aq)$ forms $CO_2(g)$, and $[CO_2(aq)]$ remains the same.

Checkpoint 4.16

1. (a) Being smaller, Mg^{2+} has a higher enthalpy of hydration.
   (b) See § 4.1.
   (c) Magnesium carbonate is more soluble than calcium carbonate.
2. (a) Formation as described in § 4.16 took place in most countries.
   (b) See § 4.16 and *ALC*, § 18.5.6.
   (c) See § 4.16.

Questions on Chapter 4

1. (a) See § 4.2.   (b) See 4.3.   (c) See 4.4.   (d) See § 4.5.
2. (a) See §§ 4.2, 4.3.
   (b) See §§ 4.4, 4.5.
   (c) Physical weathering: Water enters cracks in rocks, expands as it freezes, exerting pressure which widens the cracks. Chemical weathering: see § 7.3.
3. (a) See § 4.1 and Figure 4.1A.
   (b) See § 4.1.
   (c) See § 4.15.5. Groundwater reacts with calcium carbonate and contains carbonate ions and hydrogencarbonate ions which make it alkaline.
4. See § 4.8
5. (a) See § 4.4.13.   (b) See 4.13.1.   (c) See § 4.13.2.
6. $2(CH_2O) \longrightarrow CH_4 + CO_2$
   $60 \, g \, (CH_2O) \longrightarrow 24.0 \, dm^3 \, CH_4$ at r.t.p.
   $150 \, cm^3$ is produced by $(150/24000) \times 60 = 3.75 \, g$
7. (a) $(CH_2O) + O_2 \longrightarrow CO_2 + H_2O$
      $200\,000 \, dm^3 \times 300 \, mg \, dm^{-3} = 60 \, kg$
      With M_r of $(CH_2O) = 30 \, g \, mol^{-1}$, this mass is $2000 \, mol$
      40% is oxidised $= 800 \, mol$
      O_2 used $= 800 \, mol$
      Air used $= 5 \times 800 \, mol = 4000 \, mol$
      Air supplied $= 5 \times 4000 \, mol = 20\,000 \, mol$
      $= 20\,000 \times 24 \, dm^3 = 4.8 \times 10^5 \, dm^3$
   (b) $2(CH_2O) \longrightarrow CH_4 + CO_2$
      $2000 \, mol \, (CH_2O) \equiv 1000 \, mol \, CH_4$
      $= 2.4 \times 10^4 \, dm^3 \, CH_4$

CHAPTER 5: WATER TREATMENT

Checkpoint 5.2

1. (a) Neutralisation of excess acidity
   (b) Coagulating colloidal impurities
   (c) Killing pathogenic micro-organisms
   (d) Reducing tooth decay
2. (a) The individual has fewer dental cavities.
   (b) The Government has a lower NHS bill for dental treatment.
3. People do not have to remember to take the tablets. If the fluoride is taken in one dose, more of it is excreted than if it is taken throughout the day.
4. This avoids the formation of trihalomethanes, which may be carcinogenic; see § 5.1.

Checkpoint 5.4

1. Filtration and settling.
2. Micro-organisms, supplied with oxygen, digest organic matter in the sewage.
3. Digestion by anaerobic micro-organisms.

Checkpoint 5.7

1. (a) e.g. nitrates, phosphates, ammonium compounds, potassium compounds, humus
   (b) When it contains heavy metals
2. (a) It is slowly digested by micro-organisms.
   (b) It can drift back towards the coast.
3. (a) Ion exchange or electrolysis with added chloride ion
   (b) Activated charcoal or oxidation
   (c) Precipitate as $PbCO_3$ or co-precipitate with $Fe(OH)_3(s)$
   (d) Cation exchange
4. (a) They support the growth of algae.
   (b) Precipitate as insoluble calcium phosphate.
   (c) Nitrates are soluble.

Checkpoint 5.8

1. Mass spectrometry
2. Atomic absorption spectrometry or atomic emission spectrometry
3. Either titrate against standard acid or use a pH meter.
4. See § 4.15.3.

Questions on Chapter 5
1. (*a*) Aluminium ion
   (*b*) Hydrolysis forms a gelatinous precipitate of
      aluminium hydroxide and suspended material
      precipitates with it. In addition, the Al^{3+} ion has a
      high ratio of charge/volume and neutralises the
      charges on colloidal particles, causing coagulation.
2. See § 5.1.
3. (*a*) It contains micro-organisms; see § 5.4.
   (*b*) Micro-organisms oxidise some of the organic matter
      and convert some into biomass.
4. Filtration removes insoluble particles; reverse
   osmosis removes dissolved particles.

5. (*a*) Electrolyse with added chloride ion.
   (*b*) Precipitate as the sulphide or electrolyse or use ion
      exchange.
   (*c*) Precipitate as the carbonate or use ion exchange or
      electrolyse.
   (*d*) Precipitate as the hydroxide or carbonate or
      electrolyse or use ion exchange.
   (*e*) Precipitate as calcium phosphate.
   (*f*) Use ion exchange or electrolyse.
   (*g*) Electrolyse or use ion exchange.
   (*h*) Electrolyse.

CHAPTER 6: WATER POLLUTION

Checkpoint 6.3
1. See § 6.1. Chemically speaking, pure water is a single
   substance. Generally speaking, water is described as 'pure'
   if it is fit to drink. Unpolluted water is water that is fit to
   be used for a specified purpose.
2. See § 6.1.
3. Chlorination kills pathogens.
4. See § 4.15.3.

Checkpoint 6.4
1. See § 6.4.
2. See § 6.4.
3. See § 6.4.
4. See § 6.4.1.
5. (*a*) See §§ 6.4, 6.4.2.
   (*b*) See §§ 6.4.2. 5.7.

Checkpoint 6.8
1. See § 6.6.
2. (*a*) See § 6.7.
   (*b*) Hydrogen cyanide gas will be formed.
3. See § 6.8.

Checkpoint 6.10
1. See § 6.9.1.
2. See § 6.9.2.
3. See § 6.9.2.
4. See § 6.9.3.
5. See §§ 6.9.4, 6.9.5, 6.9.7.
6. See § 6.10.

Checkpoint 6.12
1. See § 6.11, 6.12.
2. See § 6.12.

Checkpoint 6.15
1. See § 6.13.
2. See § 6.14.
3. See § 6.9.3.
4. Pesticides sprayed on to plants pollute the air and fall to
   pollute the soil and may be leached into waterways.
   Tetraethyllead in petrol causes the emission of lead
   bromides from vehicle exhausts. The compounds pollute
   the air, fall to earth and pollute soil and crops. From the
   soil they can be leached into waterways. There are many
   other examples.

Questions on Chapter 6
1. No; see § 6.2, but the absence of coliform bacteria shows
   that the water is *not* polluted. It is easier than testing for
   pathogens.
2. They use up oxygen; see § 6.3.
3. See § 6.4.
4. See § 6.9.3.
5. See § 6.14.
6. See § 5.6.
7. See § 6.7 for cyanide, § 6.9.1 for cadmium, § 6.6 for
   sulphuric acid, § 6.8 for aluminium.
8. See § 6.9.2.
9. (*a*) Fertilisers and detergents; see § 6.4.2.
   (*b*) Fertilisers, § 6.4.1.
   (*c*) See § 6.4.

CHAPTER 7: THE LITHOSPHERE

Checkpoint 7.3
1. See § 7.2 and Figure 7.2A.
2. See § 7.2.
   (*a*) Compression
   (*b*) High pressure or temperature
   (*c*) Cooling and crystallisation
3. (*a*) Limestone is the compressed remains of the skeletons
      of marine creatures. Granite is a volcanic rock.
   (*b*) Limestone + acid gives carbon dioxide.
4. (*a*) See § 7.3.
   (*b*) (i) Reaction with acidic rain water
      (ii) Hydrolysis; see § 7.3.

Checkpoint 7.4
1. **A** clay, **B** sandy clay loam, **C** silty loam
2. (*a*) See § 7.4. (*b*) See Figure 7.4B.
3. See § 7.4.1.

Checkpoint 7.10
1. Ammonium sulphate, the salt of a strong acid with a
   weak base, is hydrolysed in solution:
   $$NH_4^+(aq) + H_2O(l) \rightleftharpoons NH_3(aq) + H_3O^+(aq)$$
2. See § 7.9.1. If cation exchange sites have been occupied by
   hydrogen ions, metal cations have been desorbed from the
   clay and can be leached from the soil.

3. See § 7.9. As cations are removed from soil by plants or by leaching, equilibria of the type
$$Ca^{2+}(clay) + 2H_2O(l) \rightleftharpoons Ca^{2+}(aq) + 2H^+(clay) + 2OH^-(aq)$$
move from left to right. The H^+ ions formed are not taken up by plants, and the pH falls.

4. See Figures 7.6B and C: montmorillonite has a larger surface area per unit mass.

5. (a) Organic humic substances possess groups such as $-CO_2H$ and $-NH_2$, which have a buffering action.
   (b) $-CO_2H$ groups and phenolic $-OH$ groups in humic substances bind metal cations.

6. The permanent cation exchange capacity arises from the adsorption of cations at the surface of silicate sheets; see Figures 7.6B and C. It is pH-dependent; see § 7.9.1. At low pH, H^+ ions displace metal cations from adsorption on clay. At high pH, OH^- ions react with clay to give a negative charge which repels cations.

7. (a) At low pH, the weathering of minerals increases; see § 7.9.2.
   (b) Hydrolysis of aluminium aqua ions yields hydrogen ions; see § 7.9.2.
   (c) When the active acidity increases, the reserve acidity increases also; see § 7.9.1.

8. First K^+ ions are adsorbed on clay silicates. When $Ca^{2+}(aq)$ is added,
   $$2K^+(clay) + Ca^{2+}(aq) \rightleftharpoons Ca^{2+}(clay) + 2K^+(aq)$$
   Potassium compounds are more soluble than calcium compounds and the equilibrium lies to the right-hand side with Ca^{2+} ions displacing many of the K^+ ions from the clay.
   When $Mg^{2+}(aq)$ is added,
   $$Ca^{2+}(clay) + Mg^{2+}(aq) \rightleftharpoons Mg^{2+}(clay) + Ca^{2+}(aq)$$
   Since magnesium compounds are more soluble than calcium compounds, the equilibrium lies over to the left-hand side, and Ca^{2+} ions are not displaced to any great extent.

Checkpoint 7.12

1. (a) The organic material in manure adds to the humus in the soil, improving the texture and allowing it to hold more air. It has a buffering action and ion exchange capacity and comes free of charge to a farmer who has livestock.

(b) Many arable farmers do not have livestock. The crops may require a different balance of N/P/K from that in manure.

2. See § 7.12.

Questions on Chapter 7

1. (a) See § 7.2.
2. (a), (b) See § 7.3.
3. (a) See § 7.2.
   (b) On addition of acid, limestone reacts to give carbon dioxide.
4. (a) Soil acidity increases.
   (b) Ammonium ions are retained at cation exchange sites.
   (c) The plants do not take up nitrates readily and the excess of nitrates is leached from the soil.
   (d) See § 6.4.
5. (a) Waterlogging
   (b) Reduction of nitrates [reverse of equation in Question 4] and reduction of metal ions to lower oxidation states with an increase in solubility and an increase in leaching from soil.
6. (a) See § 7.7. (b) See § 7.12. (c) See § 7.7.
7. (a), (b) See § 7.10. (c) See § 7.11.
8. (a) See Figure 7.4A.
   (b) See Figure 7.5C.
   (c) See § 7.6.
   (d) Clay holds more hydrogen ions than sand as reserve acidity and these require neutralising after the active acidity has been neutralised.
9. (a) See § 4.8.
   (b) $CaCO_3(s) + CO_2(aq) + H_2O(l) \rightleftharpoons Ca(HCO_3)_2(aq)$
   (c) See § 4.5.
10. (a) (i) See § 7.10. (i) See § 7.9.1.
    (b) At low pH, groups such as $-CO_2H$ and phenolic $-OH$ in organic material are not ionised. At high pH they are partially ionised as $-CO_2^-$ and $-O^-$ and bind more cations.
    (c) Ammonium ion could be lost as gaseous ammonia.
    $$NH_4^+(aq) + OH^-(aq) \longrightarrow NH_3(g) + H_2O(l)$$
11. The cations are adsorbed on ion exchange sites in the humus.
12. (a) Bacteria mediate the oxidation of NH_4^+ to NO_3^-.
    (b) The bacteria are aerobic. Waterlogged soils lack air.

CHAPTER 8: LAND POLLUTION

Checkpoint 8.4

1. (a) Not biodegradable; see § 8.4.3.
   (b) Animals at the beginning of the chain are eaten by predators, the predators are eaten by larger animals and so on; see §§ 8.4.3, 6.9.3.
   (c) A pesticide which is not broken down by micro-organisms and therefore remains in the environment for a very long time
2. (a) Organophosphates are biodegradable; see § 8.4.4.
   (b) Carbamates and pyrethrins are less toxic to human beings.
3. See § 8.4.7.
4. The groups attached to the carbon atoms of the double bond are coplanar, whereas in DDT there is free rotation about the single $C-C$ bond. The receptor site may not be able to accommodate this large planar molecule.

Checkpoint 8.6

1. Arsenic compounds > chlorphenoxy compounds > paraquat
2. (a) An alarm which sounds when the pressure rises, a back-up safety valve, etc.
   (b) The 3 Cl atoms make trichlorophenol a strong acid [see *ALC*, § 12.7.7], and it is neutralised by bases such as sodium hydroxide and sodium carbonate.
   (c) There is no 'correct' answer. The scale of the disaster was unprecedented, the experience to deal with it was lacking, and the seriousness of the damage to human health was not appreciated.
   (d) Dioxin has teratogenic effects: it affects the genes. Women who had been exposed to dioxin had deformed babies. Men who had been exposed to it fathered defective children.

Checkpoint 8.8
1. See § 8.7.
2. See § 8.8.
3. Nitrile because some hydrogen cyanide is formed.
4. **B**.

Checkpoint 8.12
1. (a) Crude oil is fractionally distilled. Some fractions are cracked with the formation of ethene, which is polymerised.
   (b) Millions of years
   (c) No
   (d) Micro-organisms do not attack them. Plastic waste rots very slowly.
2. (a) Good thermal insulator
   (b) Possibly only minutes
   (c) It becomes non-biodegradable plastic waste.
   (d) It diffuses into the atmosphere, and eventually contributes to the greenhouse effect; this effect is less serious if the gas is a hydrocarbon than if it is a CFC.
3. (a) (i) Avoids the discomfort of removal and the necessity for a visit to a doctor for removal.
        (ii) The bag dissolves in the laundry water so it does not spread infection to another batch of laundry.
   (b) Many examples, e.g. kitchenware, building materials, etc.
4. (a) Items which are in use for a short time only, e.g. carrier bags, food cartons
   (b) Many examples of items which are in use for a long time, e.g. clothing, garden chairs

5. See § 8.11.
6. (a) See § 8.10.
   (b) Different kinds of glass can be melted down together; plastics need to be separated.
7. (a) Saving Earth's reserves of metal ores, saving the energy used in extracting metals from their ores, saving on landfills for dumping metals.
   (b) Aluminium does not corrode.
   (c) (i) Use a magnet to attract the iron.
       (ii) Use a flotation tank with a liquid of suitable density so that aluminium floats and zinc sinks.

Checkpoint 8.16
1. (a) See § 8.16.1. (b) See § 8.16.2. (c) See § 8.16.3.
2. See § 8.16.1.
3. See § 8.16.4.

Questions on Chapter 8
1. See § 8.7.
2. See §§ 8.7, 8.8.
3. (a) See §§ 8.2–8.6.
   (b) Organic pollutants bond to the organic material in humus. Inorganic pollutants are held by cation exchange sites and water removes them by reactions such as
   $Sr^{2+}(clay) + 2H_2O(l) \rightleftharpoons Sr^{2+}(aq) + 2H^+(clay) + 2OH^-(aq)$
4. See § 8.6.
5. There is no 'correct answer'.
6. Leachate can more readily be collected.
7. The long half-lives of some radioisotopes; see § 8.16.

Index